Science and the Big Issues of Our Time

**The science of how our world came to be,
and our choices for the future**

By Martin Gellender

Copyright 2015 Martin Gellender

ISBN 978-0-646-93680-2

Cover photo: The Earth, as seen by Apollo astronauts orbitting the moon. Source: NASA

Science and the Big Issues of Our Time

The science of how our world came to be, and our choices for the future

Part 1: Our place in the universe

1	A brief guided tour of the solar system and beyond	1
2	Orbital motion	4
3	Orbits of Earth satellites	12
4	Prisoners of the Earth	17
5	The challenge of space exploration	22
6	Tidal forces	30
7	The birth, life and death of stars	34

Part 2. The Earth, its resources and how we use them

8	The essence of atoms	46
9	Chemical reactions	54
10	The age of metals	59
11	Aluminium & electrolytic refining	64
12	The transport revolution reshapes the world	68
13	Electrochemical batteries	74
14	More on batteries and fuel cells	79
15	Fuel cell, electric and hybrid vehicles	85
16	Petroleum fuels	90
17	Oil refining and the birth of the petrochemical age	96
18	Petrochemicals and the age of plastics	102
19	Mimicking nature with man-made materials	110

Part 3. Living on Earth

20	The Earth as a giant heat engine	116
21	Water world	125
22	Keeping cool: The science of refrigeration	131
23	Heat engines	138
24	Internal combustion engines	147
25	Feeding the world's population	154

1. A brief guided tour of the Solar System and beyond

To gain an understanding of the solar system, our immediate neighbourhood in the universe, we need to appreciate the relative sizes of the Earth, our moon and the sun. The distances involved in our solar system, and the universe beyond, is vast. To put these huge distances and sizes in perspective that we can visualise and imagine, we can construct a miniature "scale model" of the solar system, with everything shrunk in the same proportion (about 100 million times). We can view everything in the solar system relative to our Earth being a globe that is 10 cm in diameter (about the size of a grapefruit, or a small globe sold in novelty stores).

On this scale, the moon is 27 mm in diameter, smaller than a ping pong ball. The diameter of the moon is about one-quarter that of the Earth, but since the volume of a sphere varies with its diameter to the third power, the volume of our moon is roughly $(4)^3$ = 4 X 4 X 4 = 64 times less than the volume of the Earth. The mass of the moon is an even smaller fraction of the Earth's mass, since the moon has a lower density (apparently because it lacks the dense iron core present at the centres of the Earth).

On this scale, the moon is located at a distance of 3 metres from Earth. This is 30 times the diameter of the Earth.

By comparison with the Earth (or any of the other planets), the sun is huge! Not only is the sun much larger than the Earth: the mass of the sun is far more than all the planets and moons in the solar system combined. Consequently, the gravitational force exerted by the sun dominates the solar system and maintains all the planets in their orbits.

Compared to our grapefruit-size Earth, the sun would be 11 metres in diameter (about the size of a two-story house). The diameter of the sun is more than one hundred times that of the Earth (in fact, 109 times), so the volume of the sun is 1.3 **million** times that of the Earth. The orbit of the moon around the Earth would easily fit within the sun.

To maintain the correct scale, where should we put an 11-metre-diameter model of the sun if our 10 cm globe of the Earth in King George Square is in front of Brisbane City Hall? It should be 1.2 kilometres away, situated at the farthest point in the Brisbane Botanic Gardens.

How far out does the solar system extend? The Earth is one of eight planets orbiting our sun (note that a ninth planet, Pluto, has been demoted to a planetesimal). Our solar system is comprised of the sun, the planets and numerous moons orbiting the various planets.

So, far, in our miniature scale model of the universe, we have placed the sun at the far end of the Brisbane Botanic Gardens with the Earth globe in the u3a classroom (with the moon 3 metres away from the Earth). Where would we find Neptune, the furthest planet in the solar system? It should be placed 35 kilometres away, perhaps at the northern tip of the Redcliffe peninsula.

Neptune is 30 times further away from the sun than the Earth. Since the intensity of sunlight reduces with **square** of the distance from the sun, sunlight reaching Neptune would have 1/30 X 1/30 = 1/900 of the intensity of sunlight reaching the Earth. The surface of Neptune would be a very cold place: in the middle of the Neptunian day, the sun would simply look like a bright star in a black sky.

If you are surprised that the solar system is such a big, mostly empty, space, you haven't seen anything yet. Once we go beyond our solar system, the dimensions become quite mind-blowing.

Imagine holding our 10 cm scale model Earth globe in your hands. To keep the correct scale, where should we place the scale-model of the nearest star to our solar system, Alpha Centauri? We would take it nearly to the actual distance of the moon! To get to Alpha Centauri, we would need to go 10,000 times further than to Neptune at the outer fringes of the solar system.

The relative sizes of the Earth, moon and sun are summarised very well in a six minute video which you can view on-line: https://www.youtube.com/watch?v=FjCKwkJfg6Y

To summarise what we have discussed, here is a table showing the actual dimensions (in kilometres), and size relative to the Earth being 10 cm in diameter.

	Actual distance (kilometres)	"Scale model distance" relative to the Earth being 10 cm diameter
Diameter of Earth	12,700 km	10 cm
Diameter of moon	3,476 km	2.7 cm
Earth-moon distance	384,000 km	3 meters
Diameter of the sun	1.4×10^6 km	11 metres
Sun-Earth distance	150×10^6 km	1.2 kilometres
Sun-Neptune distance	4.5×10^9 km	35 kilometres
Sun to nearest star	41.5×10^{12} km	320,000 kilometres

Because the distances involved in astronomy are so mind-blowingly big, it becomes impractical and incomprehensible to express distances in terms of millions, billions or trillions of kilometres. A convenient yardstick is the time that it takes light to travel a particular distance - usually expressed in terms of "light years" (the distance travelled by light in one Earth year). The speed of light is 300,000 kilometres per second, which is much, much faster than anything in our normal experience (more than 1 million times faster than the cruising speed of an Airbus 380). The basis of Einstein's Theory of Relativity is that the speed of light is universally constant for all observers, regardless of where they are located in the universe or how fast they are travelling. The speed of light is an absolute maximum speed limit for the universe (as Einstein showed, it is one of the few things in the universe that *is* absolute and universal for all observers in all reference frames).

Here are some of the above distances expressed as the time required for light to travel each distance.

	Time for light to travel specified distance
Earth-to-moon distance	1.3 seconds
Sun-to-Earth distance	8 minutes
Sun-to-Neptune distance	4 hours
Sun to nearest star	4.3 years

By expressing distances to other bodies in light-years, this tells us how long it would take to send a signal, or to detect any change that has occurred. Astronauts visiting the moon experienced a time delay of at least 2.6 seconds before they could receive an immediate response to their communications to Earth. While this delay must have been an irritating annoyance to astronauts trying to have a normal conversation with Mission Control on Earth, think about what this would be like to an astronaut visiting Alpha Centauri (setting aside the huge issue of how he would get there). By the time the young astronaut heard that his wife had had a baby boy, his son would be nearly ready to start school.

The time delay in communication would be far, far greater for other stars near to Earth that we can see with the unaided eye. In most such cases, by the time that a message arrived that the astronaut's wife had given birth to a baby boy, his son would already have died of old age!

The vast majority of stars within our own galaxy are much too far away, and much too faint, for we Earthlings to see with the unaided eye. On a clear night, away from the lights of a city, you can expect to see about 2,000 stars of the roughly 100 **billion** stars in our galaxy. When astronomers view stars from just beyond our immediate galactic neighbourhood, using powerful telescopes, they see light that was emitted tens of thousands of years in the past.

2. Orbital motion

Gravitational force

The continued existence of our solar system, our galaxy and the entire universe depends on the force of gravity. Gravity is an attractive force which binds the Earth and other planets to the sun, and prevents them from flying off into the vast, empty coldness of space.

The force of gravity acts between all objects having any mass. Normally, we don't think of gravitational forces acting between your body and the chair or person next to you – but they do! However, in these cases, the gravitational forces are insignificant, probably at the limits of what we could measure using the most sensitive instruments.

On the other hand, gravitational forces become very significant when one or more of the objects has enormous mass, such as the Earth. The mass of the Earth is about 100,000 billion billion times greater than a person standing next to you on the bus. Consequently, the gravitational force of our bodies being pulled toward the Earth is very noticeable, and this force - our weight - affects every aspect of our lives. As we shall see, the downwards force of our weight keeps each of us, and the rest of humanity, confined near the surface of the Earth.

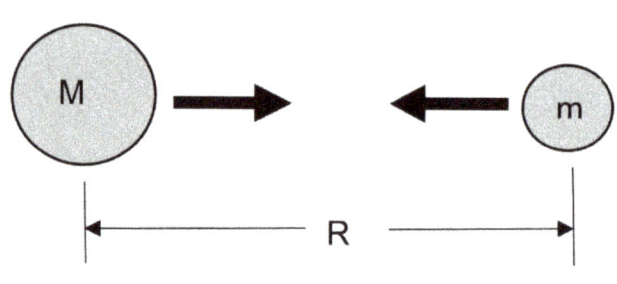

Let's consider the gravitational force acting between any two objects of mass **M** and **m** respectively, separated by distance **R**.

The force of attraction varies directly in proportion to mass **M** and to mass **m**, and is inversely proportional to the square of the distance **R** between the masses. The force of gravitational attraction is given by:

$$\text{Equation (1)} \quad \text{Gravitational force} = \frac{GMm}{R^2}$$

G is simply a constant, termed the Universal Gravitational constant, which appears to apply universally to all masses throughout the universe. The value of **G** is 6.67×10^{-11} m^3/kg-sec^2.

G is a very small number, so the force of gravity is normally insignificant unless there is a very large mass involved (like the Earth, sun, etc).

Note that, for a spherical body (like the Earth, sun and all the planets and stars), all of the mass of the body "acts" as if is located at the centre of the sphere. This is very convenient, and simplifies our calculations.

We know that many man-made satellites and the moon are orbiting around the Earth. The Earth is exerting an attractive force on these objects, so why do these satellites remain permanently in orbit (in the case of the moon, for billions of years)? Why don't they simply get pulled and crash into the Earth?

Satellites are held in a circular path around the Earth because of the attractive force of gravity. If not for this gravitational attraction, these objects would simply fly in a straight line – and escape into deep space. The attractive force of gravity is exactly that required to keep these objects in a circular path.

Imagine a hypothetical experiment in which we mount a huge cannon at the top of a mountain. If the mountain were sufficiently high, it would rise above the atmosphere, and we could ignore the effect of air resistance as the projectile left the cannon.

First, let's imagine that we fire a projectile at a velocity of, say, 100 metres per second. It would continue at this horizontal velocity, but would be accelerated vertically downwards with the acceleration of gravity. The projectile would travel in a parabolic trajectory and strike the surface at some distance downrange.

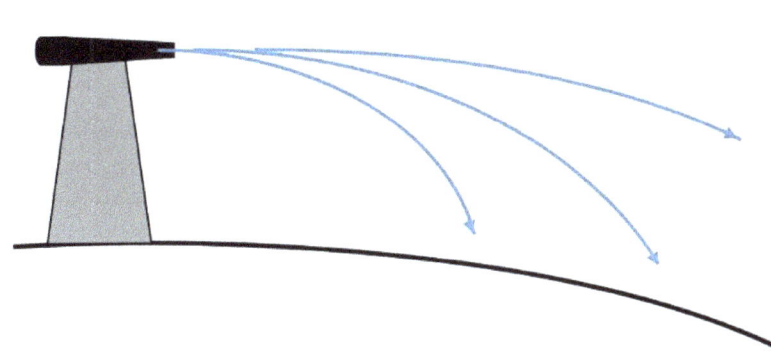

The projectile would travel further, as you might expect, if it were launched with higher velocity. With twice the muzzle velocity, the projectile would travel twice as far before it hit the ground. However, if we keep increasing the velocity, we would find that the projectile travels further than we expected. This occurs because the Earth is not flat. As the distance travelled by the projectile increases, the Earth begins to curve away from the projectile. At some very high velocity – the "orbital velocity" - we would find that the surface of the Earth curves away from the projectile at the same rate as the projectile is falling towards the Earth's surface. At this velocity, the projectile never hits the Earth! It remains at the same height above the Earth's surface, while it travels around the Earth. The projectile would be in orbit around the Earth.

It is important to realise that the Earth's gravity is still acting on a satellite in orbit around the Earth. The satellite is still falling towards the Earth, but the surface of the Earth is falling away at the same rate as the satellite is falling.

A satellite travelling in a circular orbit moves at constant speed, but its direction is constantly changing. It is constantly being accelerated towards the centre of the Earth.

Gravitational attraction, and the resulting acceleration of gravity, is "one side of the coin" that allows us to understand orbital motion. The "other side of the coin" is the acceleration that is required to maintain an object moving along a circular path. The force required to accelerate a mass towards the centre of a circular path (the so-called "centripetal force") is related to its velocity and radius, as given in Equation (2), which is derived in the next section. Those readers who are prepared to accept this result may skip the next section. Readers who are interested (or who think that I just made this up this result) might wish to read how the equation for centripetal force can be derived.

The force acting on a body in a circular path

Imagine an object moving in a circular orbital path with velocity **V** at radius **R** from the centre of a planet, star, etc. An object will move in a straight line (with no acceleration) if no force acts on it. Consequently, for the object to travel in a circular path, it must experience an inwards force which is constantly accelerating it towards the centre. Let's determine the rate of acceleration and the inwards "centripetal force" required to maintain its circular path.

Let's consider the motion of the object over a very short (infinitesimal) period of time Δt, during which it travels a very small sector of its circular path.

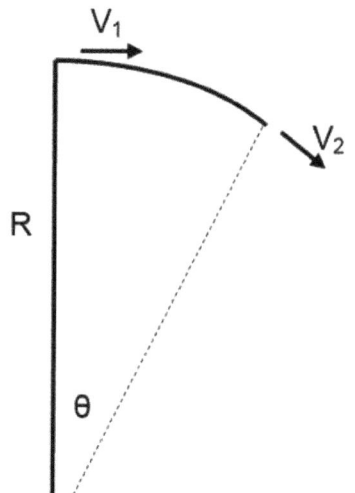

During time period **Δt**, the object will move a distance **VΔt** along the arc of its circular path.

Distance along arc = VΔt

This circular arc will subtend an angle **θ**. If angle **θ** is expressed in radians, this is the ratio between the length of the arc and the radius of the circle (with 2π radians making a full circle of 360°). Consequently, the length of the arc travelled by the object in time Δt can also be expressed in terms of angle **θ**.

Distance along arc = Rθ

We can combine these two simple equations to relate the momentary time **Δt** required for the orbiting body to subtend angle **θ** along the circular path. This gives simply:

$$V\Delta t = R\theta \quad \text{and thus,} \quad \Delta t = \frac{R\theta}{V}$$

As the object moves along the circular path, the magnitude of its velocity (its speed) doesn't change. The initial speed V_1 is exactly equal to its speed V_2 after time **Δt**. But velocity is a vector quantity that depends upon its magnitude *and* its direction. So, let's take the velocity arrow vectors V_1 and V_2 in the diagram above, and align them so that they have the same starting point. Using a bit of geometry, it is possible to show that the two vector arrows differ in their orientation by the same angle **θ**.

The change in velocity is given by the vector arrow **ΔV**, which is perpendicular to velocity vector **V**.

Now, we are considering the motion of the object over an extremely short (infinitesimal) period of time, so angle **θ** will be very small. In this case, the length of the velocity vector arrow **ΔV** is given by:

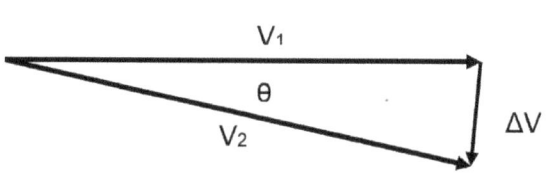

ΔV = Vθ

The acceleration that the object undergoes during its motion along its circular path is equal to the change in velocity **ΔV** divided by period of time **Δt**. So, let's determine the ratio of ΔV/Δt, in the limit as Δt becomes infinitesimally small.

$$\text{Inward acceleration of object in circular path} = \frac{\Delta V}{\Delta t} = \frac{V\theta}{R\theta/V} = \frac{V^2}{R}$$

The force required to cause this inwards acceleration is given by the mass of the object **m** times the acceleration **V²/R**.

> **Equation (2)**
>
> Force required to maintain mass m in circular path = $\dfrac{mV^2}{R}$

Orbital motion

For an object in orbit, like a planet orbiting the sun, the gravitational attractive force towards the sun provides the centripetal acceleration required to keep the object in a circular orbit. Equating the gravitational force exerted by the sun on a planet (given by Equation 1) with the centripetal force (given by Equation 2) gives:

$$\frac{GM_{sun}m_{planet}}{R^2} = \frac{m_{planet}V^2}{R}$$

Note that m_{planet}, the mass of the planet, can be cancelled out, and the equation can be simplified to give the orbital velocity V:

$$V^2 = \frac{GM_{sun}}{R} \qquad \text{So,} \qquad V_{orbital} = \sqrt{GM_{sun}/R}$$

Perhaps more interesting than the orbital velocity of a planet is the time required for it to complete one circular path around the sun, or one "year" for that planet. A planet's "orbital period" can readily be calculated by dividing the distance along its orbit (**2πR**) by its orbital velocity $V_{orbital} = \sqrt{GM_{sun}/R}$.

> **Equation 3**
>
> Orbital period = $\dfrac{2\pi R}{\sqrt{GM_{sun}/R}} = \dfrac{2\pi}{\sqrt{GM_{sun}}}R^{3/2}$

The key thing to note about Equation 3 is that the orbital period varies with the distance of the planet from the sun to the power of 3/2. So, the further a planet is from the sun, the longer is its orbital period.

This 3/2 power relationship between the distance of a planet from the sun and its orbital period was first discovered in the 17th century by the German mathematician Johannes Kepler. Kepler made this discovery by carefully studying extensive data of the positions of planets meticulously collected over many years by the Dutch astronomer Tyco Brahe.

As an example of the 3/2 power law, consider the orbit of Neptune, the outermost planet in the solar system. Neptune is 30 times further from the sun than the Earth, so the length of a year on Neptune would be $(30)^{3/2}$ times longer than on Earth, or 164 Earth years.

In fact, the discovery of Neptune in 1853 was a brilliant confirmation of the gravitational law, given in Equation (1). Being so far from Earth, Neptune is too dim to be seen with the unaided eye. As it later emerged, a number of astronomers (including Galileo) had observed Neptune since the invention of the telescope, but none had recognised it as a planet. However, by the mid-19th century, astronomers were well aware of the planet Uranus (discovered about sixty years earlier), and were puzzled by anomalies in its path. Could these anomalies be due to an irregularity in Uranus' gravitational attraction to the sun or, as astronomers Urbain Le Verrier and John Couch Adams suggested, could there be a further unknown planet disturbing the orbit of Uranus? Using Verrier's calculations, astronomer Gottfried Galle discovered Neptune exactly where Verrier had predicted it to be! The discovery of Neptune was a sensational confirmation that the gravitational attraction law given by Equation (1) drives the operation of the solar system and the orbits of the planets. Neptune's discovery in the mid-19th century might seem like a long time ago, but Neptune is only now completing the first orbit around the sun since it was discovered.

A few twists and turns

In this chapter, I have limited my discussion to circular orbits. I expect that most readers would be very familiar with circles, being the path that is equidistant from a point at the centre. The assumption of circular orbits gives simple, elegant and valid results (given by Equation 3). As it turns out, the orbits of the planets and moons in our solar system are very close to circular, as are most man-made satellites in low-Earth orbit and all satellites in geostationary orbit.

However, not all orbits are circular. In particular, some "exosolar planets" recently discovered around other stars follow orbits which are in highly eccentric elliptical orbits. So too are meteorites which strike the Earth. Meteorites originate from circular orbits in either the asteroid belt beyond Mars or in the Oort belt beyond Neptune, but their orbits are disturbed by impacts or the gravity of other passing bodies. These meteorites can pass close to the sun, be accelerated by the sun's gravitational attraction, and then fly off to the outer reaches of the solar system. They then travel in elliptical orbits, with the sun being one of the two foci points of an ellipse.

While a circle has a single central point from which all points are equally distant, an ellipse has two foci points. For an ellipse, the sum of the distances of any point from the two foci is constant. The "eccentricity" of an ellipse is the ratio of the distance between the foci to the furthest distance along the axis of the ellipse. A circle is a special case of an ellipse, where the two foci are located at the same point (thus, a circle is an ellipse with zero eccentricity).

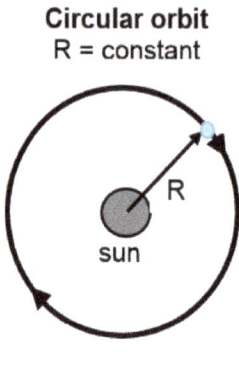
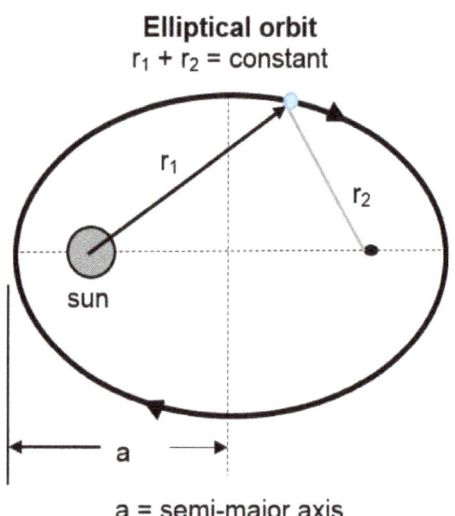

As it turns out, all orbiting objects follow an elliptical path, although in many cases, the elliptical path is nearly circular. The calculations in this chapter have been derived for the special case of circular orbits, but we can generalise our equations and results so that they apply to any object in orbit by substituting the "semi-major axis" **a** of the orbit, instead of a radius **R**. Thus, the time required for an object to complete one orbit is proportional to the 3/2 power of its semi-major axis. Of course, in the special case of a circular orbit, the radius of the orbit is equal to its semi-major axis.

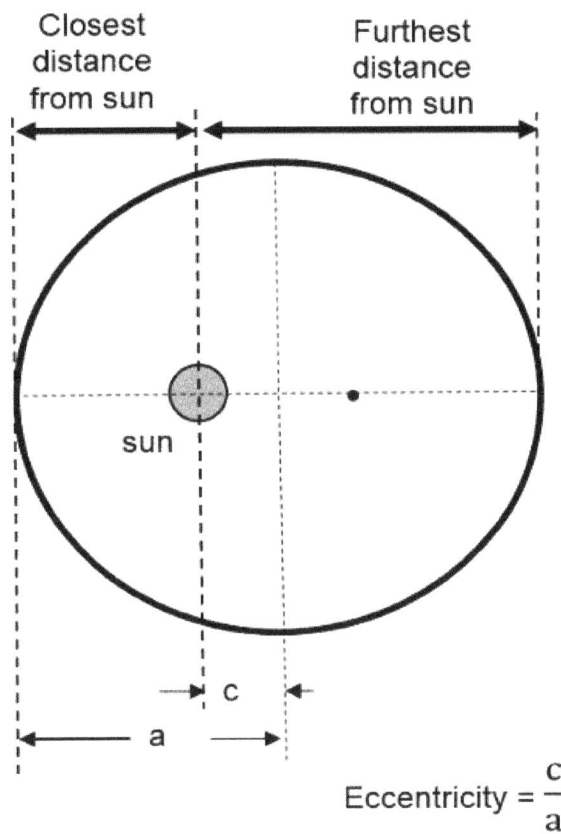

Eccentricity = $\dfrac{c}{a}$

The orbit of the Earth around the sun is nearly, but not quite, circular, with an eccentricity of 0.016. This means that, at its furthest point from the sun, the distance from the Earth to the sun is (1+0.016) times the semi-major axis of its orbit. At its closest point to the sun, the Earth is (1-0.016) times the semi-major axis. Consequently, the furthest point of the Earth from the sun is 3% further than its nearest approach. At its furthest point from the sun, the intensity of sunlight is about 6% less than at its nearest point.

The closest point of the Earth's orbit to the sun (its perihelion) occurs in mid-summer in the southern hemisphere, causing sunlight reaching the Earth's surface in Australia to be slightly stronger in summer. The more intense solar radiation slightly enhances the effect of longer days, shorter nights and more direct solar radiation due to the tilt of the Earth on its axis.

The furthest distance of the Earth from the sun (its aphelion) occurs in mid-winter in the southern hemisphere.

But exactly the opposite occurs in the northern hemisphere, where the sunlight is actually slightly *more* intense in winter! I can assure the reader, having spent the first half of my life living in the Northern Hemisphere (northern US, Canada and the UK) that the winters can be bitterly cold (far colder than I have experienced in Australia), despite the Earth being closer to the sun. Similarly, the summers in New York or Toronto can be just as hot and oppressive as they are in Brisbane, despite the sun being slightly further away.

Varying sunlight levels as the Earth moves along its elliptical orbit throughout the year play a much smaller role on the seasons and climate than does the tilt of the Earth on its axis. This should not be surprising. The tilt of the Earth at the declination angle of 23.5° causes the angle of incident sunlight to change dramatically throughout the year, with a corresponding huge variation in solar energy shining on each square metre of the Earth's surface. For a clear day in Brisbane (located at 27.5° South latitude), **twice the solar radiation** falls per square metre of horizontal surface in mid-summer than in mid-winter. The increased "insolation" in summer months is partly due to greater intensity of sunlight striking the surface, and partly due to the longer duration of daylight hours (see the graph below). The reduction in solar intensity from mid-summer to mid-winter is even more pronounced as we move away from the equator, with a 100% reduction in the Antarctic polar region (where the sun doesn't rise above the horizon in mid-winter).

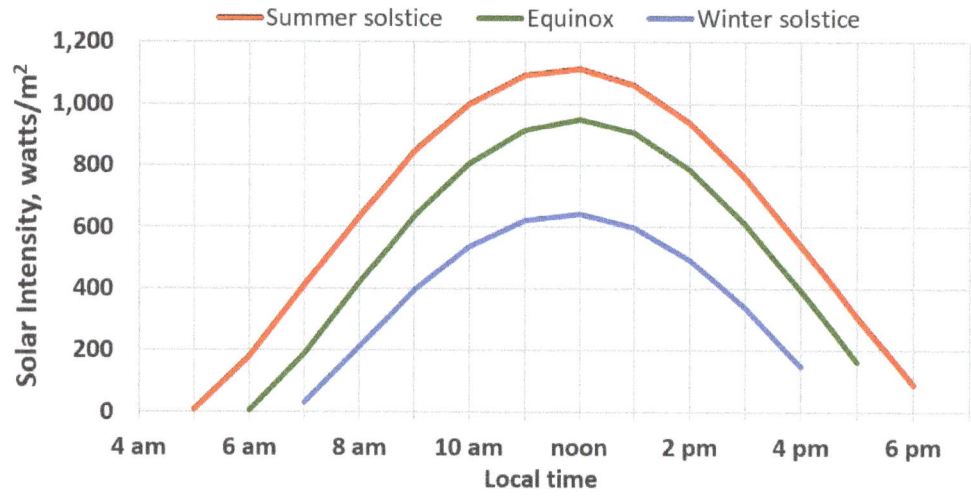

Source of data: Tables of solar position and radiation for Brisbane, by J.W. Spencer, published by CSIRO 1979, www.publish.csiro.au/books/download.cfm?ID=7199

Source of data: Tables of solar position and radiation for Brisbane, by J.W. Spencer, published by CSIRO 1979, www.publish.csiro.au/books/download.cfm?ID=7199

Now, let me turn to another surprising and unexpected complication. We have seen that the Law of Gravitation, given by Equation (1), neatly and precisely explains everything we know about the movement of planets, meteorites and man-made space satellites within our solar system. Most scientists believe that the Law of Gravitation applies universally to everything in the universe. But there is a problem. When astronomers look at distant spiral galaxies, similar to our own Milky Way Galaxy, they see billions of stars travelling in circular orbits around the centre of the galaxy (where there resides a supermassive black hole, with a mass of about a million stars). However, when they estimated the mass of all known stars and other objects (including the supermassive black hole, planets, meteorites, dust and gas), the gravitational attraction pulling stars towards the centre of the galaxy was not sufficient to balance the outwards centrifugal force produced as the galaxy spins.

A few astronomers have theorised that the Universal Gravitational Constant G is not universal at all, but perhaps it slightly increases at the much, much larger distances on the scale of spiral galaxies (whose diameter is about 100 million times that of our solar system). However, most astronomers believe that the Gravitational Constant G is indeed constant at all scales throughout the universe. They believe that spiral galaxies do not fly apart because their mass is much larger than what it appears based on what we can see through telescopes. The missing mass is called "dark matter". Nobody knows what this "dark matter" might be, except that it is not stars or other stuff that we can see. We know that "dark matter" must have mass, but aside from its gravitational effects, does not interact strongly with normal matter (otherwise, "dark matter" would have been discovered long ago). It might be sub-atomic particles which are currently unknown, and some scientists hypothesise the existence of "Weakly Interacting Massive Particles" (known as WIMPS) that fly around within galaxies. Astronomers have even mapped out the distribution that these hypothetical particles would have, and found that they are not uniformly distributed throughout the galaxy or the wider universe. In fact, it is likely that clumps of "dark matter" – whatever it is – initiated the formation of star clusters and galaxies. The gravitational attraction of these clumps of "dark matter" would have caused enormous gas clouds to contract and then to collapse into the stars we see today.

Spiral galaxies contain billions of stars orbiting around a galactic centre, containing a supermassive black hole.

Shown here is a composite photograph of the spiral galaxy known as Messier 81, taken in visible light by the Hubble space telescope, ultraviolet light by NASA's Galaxy Evolution Explorer satellite and infrared light from NASA's Spitzer satellite. Messier 81 is located about 12 million light-years away and is one of the brightest galaxies that can be seen from Earth.

Photo by: NASA/JPL-Caltech/ESA/Harvard-Smithsonian CfD

But, bear in mind that (at the time of writing) nobody has yet directly detected "dark matter" particles. It is a sobering thought that, with mankind's extraordinary technology in the 21st century, we still have no idea what makes up about 75% the mass of our galaxy!

3. Orbits of Earth satellites

In the previous chapter on Orbital motion, we saw how a force of gravitational attraction acts between any two objects (one with mass **M**, and the other with mass **m**), and that this force varies inversely with the square of the distance **R** between the two bodies. This result was given in the previous Equations (1):

$$\text{Gravitational force} = \frac{GMm}{R^2}$$

This equation is completely general, and is believed to apply universally throughout the universe.

Let us now look at the specific case of the world that we humans inhabit on Earth. We live at (or very near) the Earth's surface, at a distance of 6,350 kilometres from the centre of the Earth. In our world, the gravitational force acting on any mass **m** (which we call its "weight") is **mg**, where g is the acceleration due to gravity at the surface of the Earth (9.8 meters/second^2). Substituting this information into the general equation (above) gives:

$$\text{Gravitational force at surface of Earth} = \frac{GM_E m}{R_E^{\ 2}} = mg$$

Where R_E is the radius of the Earth (6,350 km, or 6.35 X 10^6 meters)
        M_E is the mass of the Earth (5.97 X 10^{24} kilograms, if you are interested),

Note that the mass **m** of the object appears on both sides of the equation and cancels out, so that:

$$GM_E = R_E^{\ 2}\, g$$

This allows us to rewrite an equation for the *specific case* of the gravitational force between the Earth and any body of mass **m** at distance R from the centre of the Earth.

> Gravitational force on an object at distance R from the Earth = $\left[\dfrac{R_E}{R}\right]^2 mg$
>
> Note that **mg** is simply the weight of mass **m** at the surface of the Earth.

In the previous chapter, we also derived general equations relating the velocity of any object in a circular orbit around a planet or star (with mass **M**), and the time for the object to complete a circular orbit of radius **R**. These results were completely general for any planet, moon or man-made satellite in orbit around a more massive body.

$$\text{Orbital velocity} = \sqrt{GM/R}$$

$$\text{Orbital period} = \frac{2\pi}{\sqrt{GM}} R^{3/2}$$

Once again, we can modify these equations for the specific case of a satellite in orbit around the Earth. What we find is:

> Orbital velocity, for satellite in orbit around Earth = $\sqrt{gR_E^2/R}$
>
> Orbital period, for satellite in orbit around Earth = $\dfrac{2\pi}{R_E \, g^{1/2}} R^{3/2}$

Satellites in low-Earth orbit

Most man-made satellites are in "low-Earth orbit", only a few hundred kilometres above the Earth's surface. This is above the relatively thin layer of air that blankets the Earth (which we call the "atmosphere"). Consequently, satellites experience little or no air resistance at this height and can remain in orbit for many years. In the case of low-Earth orbit, distance **R** is only slightly larger than the radius of the Earth R_E, and the acceleration of gravity **g** has virtually the same value as at the surface of the Earth. It is quite easy to show by mathematical substitution that:

Orbital velocity (low-Earth orbit) = $\sqrt{g\, R_E}$

Orbital period (low-Earth orbit) = $2\pi \sqrt{\dfrac{R_E}{g}}$

Since R_E, the radius of the Earth, is 6,350 kilometres (6.35 X 10^6 metres) and **g** is 9.8 metres/sec^2, this gives an orbital velocity of 8,000 meters/second. We can readily calculate that the orbital period for satellites in low-Earth orbit is about 5,000 seconds (or slightly longer, about 90 minutes, for a satellite in orbit several hundred kilometres above the Earth's surface).

About 500 satellites are currently operating in low-Earth orbit. Notable among these are the International Space Station, the Hubble space telescope and many Earth observational satellites. In addition, there are thousands of defunct satellites, rocket booster stages and small pieces of other "space junk" (the odd astronaut's glove or loose bolts that drifted away during space missions).

Orbiting space junk is a serious threat to operating satellites. Although many of these pieces are quite small, they would likely cross the orbits of other low-Earth satellites at relative speeds of thousands of kilometres per hour. Collisions with space junk would not only destroy operational satellites, but could tear a satellite apart and produce a shower of even more debris. Such a disaster is graphically depicted in the 2013 movie "Gravity".

Satellites in polar orbit

Many satellites launched into space are placed into polar orbits, circling the Earth in a north-south plane, travelling over the north and south poles. As they circle the Earth, the Earth rotates below the satellite. During each orbit, a new swath of the Earth's surface comes into view underneath the path of the satellite. This allows one satellite to photograph or take measurements of the entire surface of the Earth over the course of each day. Such polar orbits are widely used for military spy satellites and for many different types of surveillance and remote monitoring (measuring changes in vegetation, ocean temperature, etc).

During each rotation of the satellite, requiring about 90 minutes (or 1/16th of a 24 hour day), the Earth will complete 1/16th of one rotation. At the equator, the Earth will rotate 1/16th of its 40,000 kilometre circumference – or about 2,400 kilometres – during the time that the satellite takes to circle the Earth from one orbit to the next.

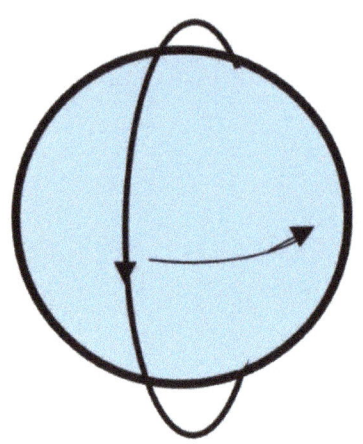

For the satellite to "see" the entire surface of the Earth each day, without any gaps, it must be able to see 1,200 kilometres to the horizon on the east, and 1,200 km to the horizon on the west, as it crosses the equator.

We can calculate the minimum height **H** that the satellite must orbit above the Earth if it is to have clear view to the horizon 1,200 kilometres away.

Let's draw a simple diagram (not to scale), showing a satellite at distance **H** above the Earth's surface, and therefore, at distance R_E+H from the centre of the Earth. The satellite has a view to the horizon of distance D. The line from the satellite to the horizon is tangent (parallel) to the Earth's surface, and forms one leg of a right triangle, as shown in the figure.

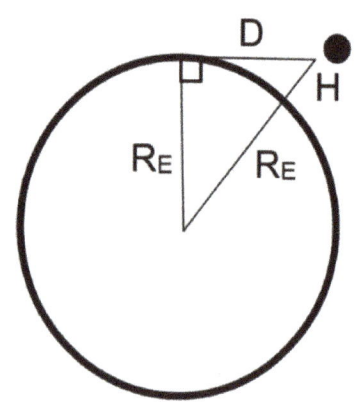

One side of the triangle is the line-of-sight distance **D** from the satellite to the horizon. The other side of the triangle goes from the centre of the Earth to the horizon point: its length is simply the radius R_E of the Earth. The hypotenuse of the triangle goes from the centre of the Earth to the satellite at height **H** above the Earth's surface. The length of the hypotenuse is (R_E+H).

The Pythagorean Theorem relates the length of the hypotenuse of a right triangle to the length of its two sides. It states that the sum of the squares of the two sides of the triangle is equal to the square of the hypotenuse.

Applying the Pythagorean Theorem to the triangle above gives:

$$R_E^2 + D^2 = (R_E+H)^2$$

Multiplying out the $(R_E+H)^2$ term gives:

$$R_E^2 + D^2 = R_E^2 + 2R_EH + H^2$$

We can cancel out the R^2 terms which appear on both sides of the equation. This gives:

$$D^2 = +2R_EH + H^2$$

We can simplify the equation further by a simple approximation. In low-Earth orbit, or other situations near the surface of the Earth, the height **H** above the surface of the Earth is much, much less than the radius of the Earth R_E. This means that the term H^2 will be insignificant in comparison with the $2R_EH$ term, and this term can be ignored without affecting the accuracy of the result. When we do this simplification, we get:

$$D^2 = 2R_EH, \quad \text{so} \quad D = \sqrt{2R_EH}$$

This equation is pretty nifty. It tells us the distance **D** to the horizon from any height **H** above the surface of the Earth.

Recall that, for a satellite in polar orbit to view the entire surface of the Earth, it must be able to see 1,200 kilometres to the horizon to the east, and 1,200 km to the west, as it crosses the equator. So, we need to know at what height **H** above the Earth's surface the satellite must travel to see a distance **D** to the horizon of 1,200 kilometres. Since the radius R_E of the Earth is 6,350 kilometres, we can solve the equation to find that the satellite must orbit at least 114 kilometres above the Earth's surface.

This is actually a convenient result. If we want the satellite to remain in orbit for a long time, it must be at a height of at least 100 kilometres so that the satellite orbits well above the Earth's atmosphere. Satellites in lower orbits are gradually slowed down by air resistance, lose altitude and eventually spiral down to Earth.

Geostationary orbits

Satellites used for communications and Global Positioning Systems (GPS) have entirely different requirements. For this application, we need the satellite to be remain at a fixed position in the sky relative to each location on the ground. Of course, the satellite will be moving as it orbits the Earth, but the surface of the Earth will also be moving as the Earth rotates. The "trick" is to place the satellite in a circular orbit around the equator, at a height where its orbital period matches the 24 hour rotational period of the Earth. From a point on the Earth's surface, the satellite appears to remain at a fixed position in the sky. Such an orbit is known as a "geostationary orbit".

It is clear that such satellites would not be in low-Earth orbit, as the orbital period in low-Earth orbit is only 90 minutes. The further away from Earth that the satellite orbits, the greater will be its orbital period. So, we'll need to move the satellite to a very high orbit (far from the Earth's surface) so that it requires 24 hours to complete each orbit. We can use the equation derived earlier for the orbital period of a satellite to determine that a geostationary satellite must orbit at a distance of 35,785 kilometres above the Earth's surface at the equator (42,130 km from the centre of the Earth). This distance is about six times the radius of the Earth, and one tenth the distance to the moon.

For satellites at distances beyond geostationary orbit, their orbital period is greater than the 24 hour rotational period of the Earth. One notable example is the moon, which orbits at a distance of about 60 times the radius of the Earth. The moon takes about a month to complete one orbit around the Earth.

A series of geostationary satellites are spaced along the circular orbit at 35,785 kilometres above the equator. Currently, there are about 450 satellites operating in geostationary orbit. These satellites are a critical piece of the technological infrastructure upon which our modern society is based. GPS satellites send out signals which are exactly synchronised. The time difference between the arrival of two signals at any point on the Earth gives the difference in distance to the two geostationary satellites. A GPS unit receiving signals from three geostationary satellites can use triangulation to determine its exact position. This is not only useful to guide you across Brisbane through the traffic: it is used by farmers to guide their tractors and harvesters in perfect straight lines across their fields; it is used by military forces to target aircraft and missiles; it is used to determine how many zones you have travelled on the bus when you touch off on your GoCard.

The GPS in your car can tell you, within a few metres or less, where you are – anywhere on the Earth – or *almost* anywhere. Although these satellites orbit above the equator, collectively they can "see" most of the Earth's surface because they are so far away. But they cannot "see" the polar region, which is below their horizon. Consider the diagram below, which shows the distances approximately to scale.

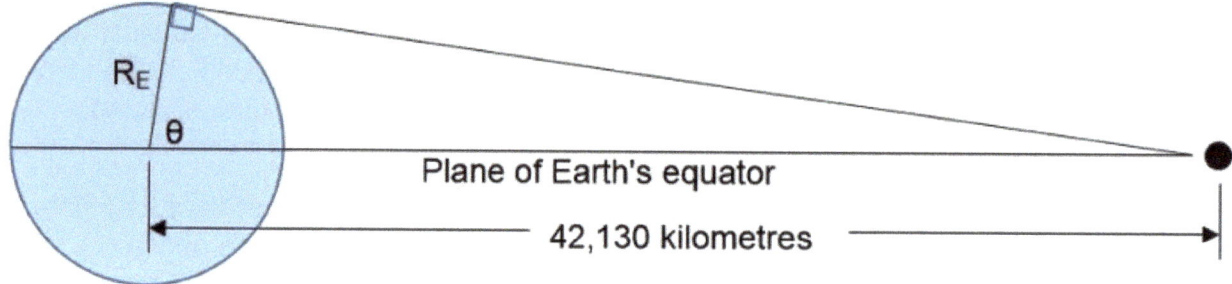

From its position within the plane of the Earth's equator, a satellite in geostationary orbit can see to the horizon on the Earth's surface at a maximum latitude angle **θ**.

The maximum latitude angle θ can be found as follows, using high school geometry:

$$\text{Cosine } \theta = \frac{6{,}350 \text{ km}}{42{,}130 \text{ km}} = 0.15 \qquad \text{So, } \theta = 81°$$

The limitation to 81° latitude is not a major problem, as there is no permanent human habitation above this latitude. If there was, these polar inhabitants could not maintain communications via geostationary satellites (and thus, could not maintain communication in "real time", that is, without transmitting to a satellite in polar orbit, with the message stored and retransmitted later).

Only 1.2% (the fraction [1-cosine 9°]) of the Earth's surface area is beyond the horizon of geostationary satellites. Even researchers working at the fringes of the Antarctic continent can speak directly via satellite with their families at home. McMurdo Station, the main Australian base in Antarctica, is located at 77.85 degrees latitude – just within the horizon of geostationary satellites!

It is not an exaggeration to say that satellites in polar and geostationary orbit are revolutionizing our lives in many ways. We (as a society) have become very dependent on such technology. However, it has become apparent that satellite technology is vulnerable, either to deliberate destruction by advanced military powers, to intense bursts of solar radiation (severe "solar storms") every few decades or centuries, and to collisions with orbiting space junk. Since advanced military forces rely extensively on satellites for their intelligence operations, communications and targeting, satellites could be a likely target in the event of a war between major military powers. Satellites are very difficult to protect, as their orbits are known and completely predictable. The Chinese government was widely condemned when they conducted a test of their anti-satellite technology by destroying one of their own satellites (spraying space junk everywhere, posing a long-term threat to other satellites). Probably the greatest inhibition for countries to destroy the satellites of potential enemies is that their own satellites would be just as vulnerable to retaliatory attack.

4. Prisoners of the Earth

All of us Earthlings are pulled down towards the Earth, and confined near the surface by the Earth's gravitational force on our mass. You'll recall that the downwards force of the Earth's gravity varies with the mass of the Earth M_E, our mass m, and is inversely related to the square of our distance from the centre of the Earth.

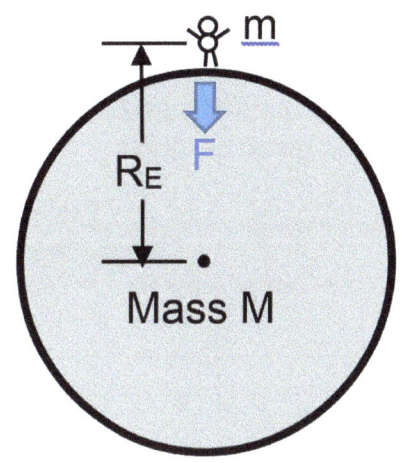

Equation (1)

$$\text{Gravitational force } F = \frac{GMm}{R_E^2}$$

At the surface of the Earth, our weight is proportional to our mass m, and to the acceleration of gravity, which is 9.8 metres/second². Bear in mind that our mass is measured in units of kilograms, but like all forces, our weight is in units of Newtons.

Weight = mg   Where g = 9.8 metres/sec²

As we venture away from the Earth's surface, our weight would diminish according to the inverse-square law, as shown in the graph below.

So, how much work is required to lift an object, say a rock with a weight of one Newton, from the surface of the Earth? The work expended is equal to the *force* that is exerted *times the distance* that the force is exerted.

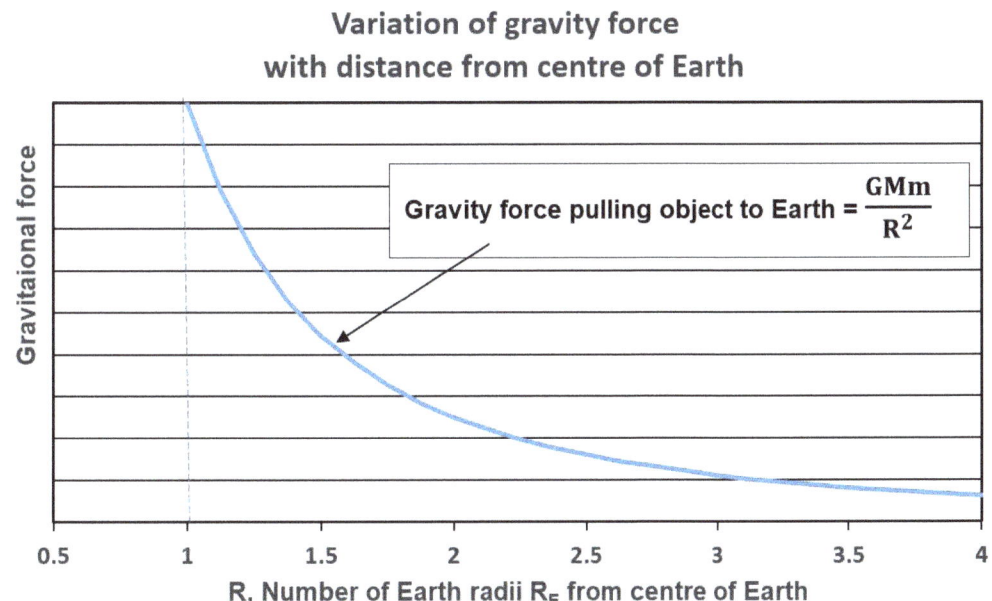

Work = (Force, Newtons)(Distance along which force is exerted, metres)

So, to lift a rock weighing one Newton to a height of 1,000 metres would require 1,000 Joules of work. To lift the rock another 1,000 metres would require another 1,000 Joules of work. We could keep lifting the rock by 1,000 metres at a time, although eventually we would be so far from the Earth that its gravitational force (and thus, the weight of the rock) would have reduced to, say, 0.99 Newton. At that point, the work required to lift the rock a further 1,000 metres would be 990 Joules. We could keep repeating this process, lifting the rock by 1,000 metres at a time, until we have lifted the rock to infinite distance, at which the Earth's gravity has faded to zero. The total amount of work required to do this is simply the total sum of the work expended each time that we lifted the rock another 1,000 metres.

This concept is depicted in the diagram below, where each rectangle represents the work required to lift the object some relatively small distance **ΔR**. The area of each rectangle is the weight of the object times distance **ΔR**. If we choose distance **ΔR** to be very small, then the sum of the areas of all the rectangles (the total work done in lifting the mass by distance **ΔR** each successive time) is the total area under the curve. In fact, the work required to lift the object from the surface of the Earth to an infinite distance is simply the area under the Force versus distance curve.

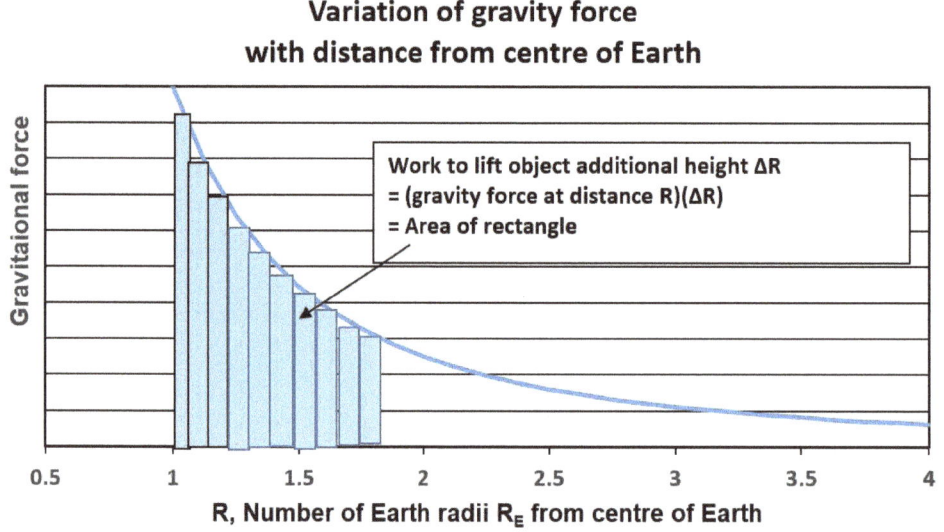

We can find the area under the Force-versus-distance curve by using the technique of "integration". For anyone who has taken 1st year calculus, this is a very simple calculation.

The result is:

Equation (2A)

Work to move mass **m** from surface to infinite distance = $\dfrac{GMm}{R_E}$

This is the "escape energy" that must be imparted for an object of mass **m** at the surface of Earth to break free of the Earth's gravity. We have seen in the previous chapter that GM_Em can be expressed in terms of the radius of the Earth **R_E** and the acceleration of gravity at the surface **g**. This substitution gives us the result:

Equation (2B)

Work to remove mass m from Earth's surface to infinite distance = $(mg)R_E$

This equation gives a simple, easily-understood result. The term **mg** is simply the weight of the object at the surface of the Earth. So, the escape energy is equal to the energy that would be required to lift the normal weight of an object (at the surface of the Earth) to a height equal to the radius of the Earth (6,350 kilometres).

So, how much energy would be required to launch a satellite into deep space? Imagine the work required to lift the satellite to the top of Mount Everest, a height of 10 kilometres. Just multiply that by 635 times.

The escape energy from the Earth is 63.5 million Joules per kilogram (63.5 MJ/kg). To put that in some perspective, this is about ten times the energy released during the explosion of each kilogram of TNT or dynamite. On a gram-for-gram basis, the escape energy is 120 times the kinetic energy of a high-velocity rifle bullet.

To launch spacecraft from the surface of the Earth requires the use of rockets, and these are not very efficient. Their actual energy consumption is much more than the escape energy of their payload. The Saturn V rocket that sent the Apollo missions to the moon converted only about 7% of the rocket's fuel energy into kinetic energy of the spacecraft. The Saturn V used about 3,000 tonnes of fuel and oxidiser to send a 50 tonne lunar lander to the moon (and most of the mass of the lunar lander was fuel for the return journey).

An asteroid or comet approaching Earth from deep space is accelerated by the gravity of Earth, and **gains** kinetic energy equal to the "escape energy" (in addition to any kinetic energy that it already had). Even after being slowed by air resistance as it passes through the atmosphere, an asteroid or comet that is large enough to reach the Earth's surface (without burning up in the atmosphere) often impacts with sufficient energy to vaporise the asteroid and surrounding rock.

The escape energy is often expressed in terms of the "escape velocity" that must be imparted to an object at the surface of a planet for it to escape the gravitational attraction of the planet. The kinetic energy of a moving object varies directly in relation to its mass **m** and its velocity **v** *squared*, as follows:

Equation (3)   Kinetic energy of moving body = ½ mv^2

Equating the kinetic energy of a body with the escape energy (from Equation 2A) gives:

$$\tfrac{1}{2} m v_{esc}^2 = \frac{GMm}{R}$$

Re-arranging to solve for the escape velocity gives:

$$V_{escape} = \sqrt{\frac{2GM}{R}}$$

Where **M** is the mass of the planet

R is the radius of the planet

If we were to fire a projectile directly upwards with a velocity that is less than the escape velocity, it would eventually fall back to Earth. An object launched with a velocity more than the escape velocity would never return to Earth (unless of course, some effect occurred to slow the object or turn it around).

This is exactly what NASA or the European Space Agency do when they launch a satellite into deep space. The rocket engines burn for only about 10 minutes. During this time, the satellite has only been propelled a few thousand kilometers from the surface of the Earth, but it has been accelerated to the Earth's escape velocity of 40,300 kilometres per hour. From then on, its velocity is sufficient to carry the satellite into deep space.

The escape energy and escape velocity are greatest for planets, moons and stars with larger masses and smaller radii. The escape velocity from a black hole at its "event horizon" is the speed of light, and the escape energy is infinite.

The table below gives the escape energy and escape velocity for a few bodies in our solar system:

	Escape Energy Megajoules/kilogram	Escape velocity
Earth	63.5	11.2 kilometres/second (40,300 km/hour)
Moon	2.93	2.4 kilometres/second
Mars	12.7	5.0 kilometres/second
Sun (from a position in Earth's orbit)	890	42.1 km/second

We Earthlings live at the bottom of a gravitational potential well of the Earth, and the energy required to climb out of this well is equal to the escape energy. Bear in mind, however, that the Earth's gravitational potential well lies within the gravitational potential well of our sun, whose mass **M** is 330,000 times greater than the mass of the Earth. A huge amount of energy would be required to move an object from the surface of the sun (which, of course, would be vaporised at the 6,000 degree temperature at the sun's surface). However, even at our position on Earth, located 150 million kilometres from the sun, the energy required to escape from our solar system (and away from the gravitational force of the sun) is 14 times the escape energy from the Earth.

Many asteroids and comets are thought to originate in the Kuiper Belt, on the outskirts of our solar system, beyond the orbits of Neptune and Pluto. When the orbits of these asteroids and comets are disturbed, they wander into the inner solar system, where they are accelerated by the gravity of the sun and the Earth. As they move inwards through the solar system and approach the Earth, they will **gain** the escape energy of the sun (at the Earth's orbit) **and** the escape energy of the Earth (in addition to any kinetic energy that they originally had within the outer solar system). Should such a comet strike the Earth, it will hit the Earth's atmosphere with a velocity of at least 44 kilometres/second (158,000 kilometres/hour). Even for relatively small comets, the energy released upon impact is huge.

During a visit to the USA a few years ago, I visited Meteor Crater in Arizona (http://www.meteorcrater.com/). This is believed to have been caused by the impact of a meteor only 30 metres in diameter. The impact crater, more than one kilometre across, is very striking, even after 50,000 years have elapsed. Yet, this was a tiddler compared to the asteroid which is believed to have killed off the dinosaurs and many other animal species. The asteroid that caused the great extinction 65 million years ago was about 10 kilometres in diameter, with a mass about **30 million** times greater than the one that caused Meteor Crater in Arizona.

In our earlier discussion on orbits, we determined the velocity of a satellite in low-Earth orbit. As it turns out, the kinetic energy of a satellite in low-Earth orbit is exactly half the escape energy. Consequently, it is much easier (half as difficult) to place a satellite in low Earth orbit than to send it to deep space. Consequently, the cost of placing satellites in low-Earth orbit is perhaps half as much as the cost of deep space missions (for a satellite of the same mass)

Recently, a number of entrepreneurs (including Richard Branson) have been promoting "space tourism". They are developing spacecraft to take wealthy "space tourists" to an altitude of about 100 kilometres on a short flight. This is a short and easy version of spaceflight, since the energy required simply to raise a spacecraft to an altitude of 100 kilometres is a tiny fraction of the kinetic energy required to place a satellite into low-Earth orbit or into deep space. The energy required to reach 100 kilometres altitude is only the weight of the spacecraft (mg) times the distance of 100,000 metres. On the other hand, the kinetic energy required to reach deep space is the weight of the spacecraft (mg) times the radius of the Earth (6,350,000 metres). In other words, to raise a spacecraft to an altitude of 100 kilometres requires only 3% of the energy required to place it in low-Earth orbit, or 1.5% of the escape energy of the Earth.

Satellites launched into the outer solar system and beyond (such as the Voyager and Pioneer satellites launched in the 1970s) must have sufficient energy to overcome the gravitational attraction of the Earth **and** the even greater gravity of the sun. For these satellites, space agencies use a trick called the "slingshot effect" (http://www.schoolphysics.co.uk/age14-16/Astronomy/text/Slingshot_/index.html). The satellite is projected on a path that nearly collides with Jupiter or one of the other planets or moons. Of course, this planet is moving within its orbit at very high velocity, so the planet acts like a cricket bat hitting a ball (although the satellite swings behind the planet, and is then flung back in the direction from which it came, rather than directly hitting and rebounding from the planet). In the process, the satellite "steals" kinetic energy from the planet (which changes its orbit to an infinitesimal degree). Some satellites are sent on course to pass near two or three planets/moons, and are accelerated in turn by the "slingshot effect" each time - a bit like a ball in the pinball machines of the 1950s.

5. The challenge of space exploration: Why launching spacecraft is difficult and expensive

We have previously seen that spacecraft in low-Earth orbit, just above the Earth's atmosphere, must be moving at a velocity of about 8,000 meters/second (28,800 kilometers per hour). To launch each kilogram of payload into orbit, we must impart 31.7 megajoules of kinetic energy. And, if we wanted to launch a spacecraft into deep space, we must impart twice as much energy, 63.5 megajoules, for each kilogram of payload. The spacecraft would be travelling at least eight times faster than a high-velocity rifle bullet, and for each unit of mass, would have 60 times as much kinetic energy. So, how can we possibly accelerate spacecraft to such speeds?

The first dreams of space travel

Interest in travel beyond the gravitational field of the Earth was first stimulated in the public imagination by publication of the book "From the Earth to the Moon" by the science fiction writer Jules Verne. In 1873. Verne proposed that the crew would be fired into space by a huge cannon. As for any projectile fired by a cannon, the spaceship and its occupants would be accelerated along the barrel by high-pressure gases produced by burning gunpowder.

A space vehicle traveling to the moon would need to attain escape velocity of 11,200 meters/second. One obvious problem with accelerating a manned space capsule to such speed by firing it from a cannon is that the rapid acceleration would squash a human being into a blob of jelly. This is not a fundamental objection because, in principle, it would be possible to build a cannon with a very long barrel (much longer than Verne envisaged) so that the astronauts could withstand the acceleration required.

If we limit the acceleration of the spacecraft to 10 g's (10 times the acceleration of gravity at the surface of the Earth), the required length of the gun barrel is 630 kilometers (one-tenth the radius of the Earth), which of course, would be completely impractical.

However, there is an even more fundamental problem with Jules Verne's premise. At the time that Verne wrote "From the Earth to the Moon", "black powder" was the standard propellant used in guns and cannons. It consisted of a mixture of finely-powdered mixture of potassium nitrate, charcoal (carbon) and sulfur. When ignited, "black powder" decomposes to produce large amounts of nitrogen and carbon monoxide gases. The reaction also releases a lot of energy (2 million Joules per kilogram of black powder mixture), heating the exhaust gases to about 2,000°C. The hot, high-pressure gases push the projectile forward along the cannon barrel and, at the same time, push the cannon barrel backwards.

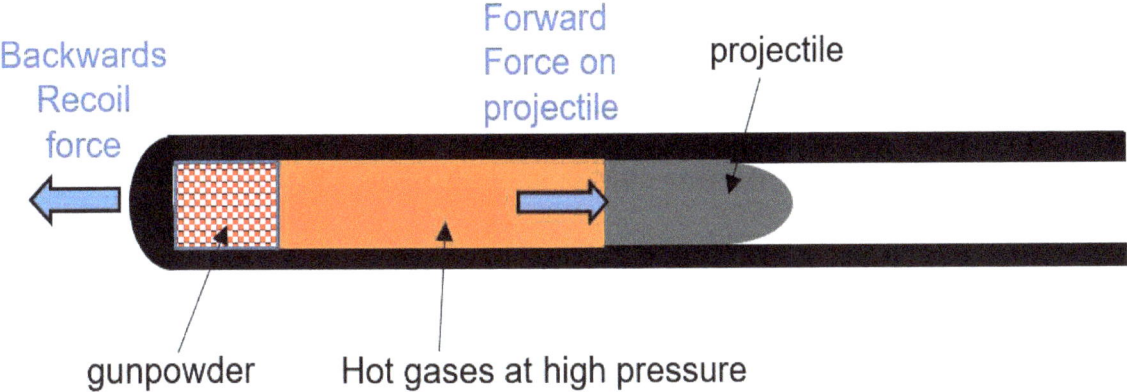

Reaction for the explosion of "black powder" propellant

$$KNO_3 + 2C \rightarrow \tfrac{1}{2} K_2CO_3 + \tfrac{1}{2} N_2 + 1\tfrac{1}{2} CO$$

Potassium Nitrate   Carbon   Potassium Carbonate   Nitrogen   Carbon monoxide

One characteristic of "black powder" is that it produces voluminous clouds of white smoke, which can be seen emanating from muskets and cannons in movies recreating battles scenes from the 17th or 18th century. The billowy white clouds arise from a fine mist of potassium carbonate droplets produced by the reaction.

Inside the barrel of a cannon, the hot, high pressure gas produced by the reaction expands and cools, doing work in pushing the projectile down the barrel. However, in accelerating the projectile to high velocity, the gas must also accelerate its own mass to high velocity. Eventually, the projectile might reach a speed at which all of the heat energy released by the reaction is converted into kinetic energy of the reaction gases. Then, no further energy would be available to further accelerate the gas or projectile.

So unless the gunpowder is accelerated along with the projectile, it would be *impossible* to fire a projectile from a cannon to escape velocity, no matter how big we made the cannon!

To find the maximum velocity to which the gases (or the projectile) can be accelerated, V_{max}, we equate the energy released by each kilogram of black gunpowder with the kinetic energy of the gas, given by $\tfrac{1}{2} m V_{max}^2$:

$$2 \times 10^6 \text{ Joules} = \tfrac{1}{2} (1 \text{ kilogram}) V_{max}^2, \quad \text{so } V_{max} = 2{,}000 \text{ metres/second}$$

This is the maximum possible velocity that we could ever hope to fire a projectile in a cannon using "black powder" as a propellant. While that is more than twice as fast as a high-velocity bullet fired by a modern rifle, it is nowhere near escape velocity! Even modern "smokeless" gunpowder (composed mainly of cellulose nitrate), first introduced in 1888 by Alfred Nobel (originator of the Nobel Prize) does not produce enough specific energy (energy per unit mass) to accelerate a projectile to escape velocity, or even fast enough to reach low-Earth orbit.

If Jules Verne wanted his novel "From the Earth to the Moon" to depict a real space journey of the future, he could not have used a cannon. Gases produced by burning gunpowder could not "catch up" with the spacecraft if it was moving at speeds approaching escape velocity.

Jules Verne should have used a propulsion technology where the spacecraft carried its propellant with it. This way, the high-velocity exhaust gases produced by the reaction do not need to travel at escape velocity. That's why when space exploration became a serious endeavour in the mid 20th century, rocket technology was used.

The modern rocket motor

A rocket carries its propellant fuel with it. The chemical reaction of the fuel generates hot gas inside the combustion chamber of the rocket. The hot gas is accelerated through a rocket nozzle, converting the heat energy of the combustion gases into kinetic energy of the exhaust gases.

Many people are surprised that a rocket engine can work in a vacuum, where there is no air or anything else to "push against". Actually, rockets "push against" their own exhaust gases. Rockets work best in the vacuum of space.

One way to visualise the force produced by the rocket motor is to consider the high pressure of the hot combustion gases pushing against the entire inside surface of the combustion chamber - except of course, the area of the nozzle (where there is no surface for the gases to push against). This creates a net force, or thrust, on the combustion chamber wall opposite to the nozzle, pushing the rocket forward.

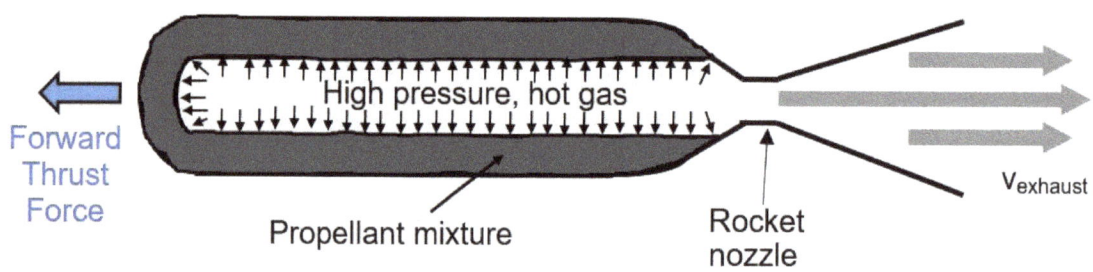

The high pressures accelerating the combustion gases to $V_{exhaust}$ will also act in the opposite direction on the rocket. Since the total momentum is conserved, forward momentum imparted to the rocket (the mass of the rocket multiplied by its change in velocity) must be equal in magnitude, and opposite in direction, to the momentum of the exhaust gases.

Based solely on the conservation of momentum it is possible to derive a "rocket equation", which relates the velocity imparted to the rocket to the exhaust velocity of the propellant gases $V_{exhaust}$ and the percentage of the rocket mass which must be propellant.

A rocket engine works exactly like a cannon, but in reverse. We can think of the hot exhaust gases as containing many small parcels of gas (projectiles) that are "fired" backwards through the rocket nozzle. This "recoil" force pushes the rocket motor forward. You may be thinking that it is "only" gas that emerges from the nozzle of the rocket engine, but don't be fooled. High-pressure gas moving at extremely high velocity carries a huge amount of momentum and energy.

Consider a small parcel of exhaust gas of mass Δm which has been accelerated to exhaust velocity $V_{exhaust}$. The backwards momentum imparted to the parcel of exhaust gas is equal to its mass Δm multiplied by exhaust velocity $V_{exhaust}$. This is equal to the forward momentum imparted to the rocket (equal to the rocket's total mass m multiplied by the additional velocity ΔV gained by the rocket).

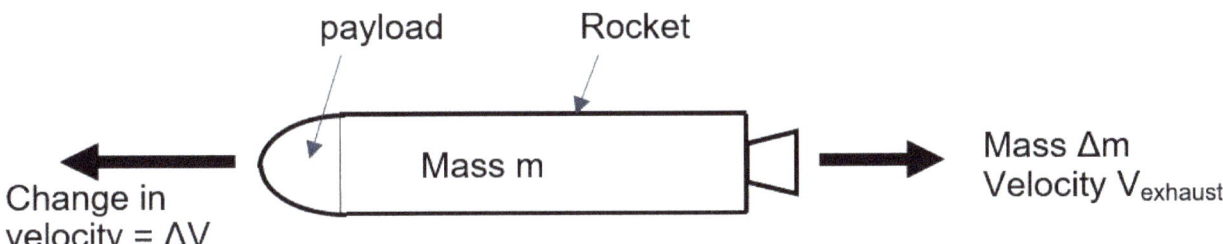

Rocket Equation: $m \Delta V = \Delta m \, V_{exhaust}$

Note that the total mass of the rocket, **m**, includes the mass of propellant carried along with the rocket. The mass of the rocket is continuously getting less as the propellant reacts and its exhaust gases are expelled at the exhaust velocity through the rocket nozzle. The rocket begins with a "launch mass", which includes the mass of propellant, the payload and the rocket itself – propellant tanks, combustion chamber, nozzle, pumps, etc. At the conclusion of its flight, all the propellant has reacted and been expelled from the nozzle, and the rocket's final mass comprises only the mass of the payload and the rocket itself.

The Rocket Equation can readily be solved by any student of 1st year calculus to determine the minimum mass of propellant required to accelerate a rocket of mass m_{rocket} to its final velocity V_{final}. Note that the mass of the rocket includes all parts of the rocket *except* the propellant (that is, the payload, fuel and oxidizer tanks, rocket chamber and nozzle, pumps, structural support and outer casing).

Mass of propellant (kg), $m_{propellant} = m_{rocket} \left[e^{V_{final}/V_{exhaust}} - 1 \right]$

What this equation tells us is that the amount of propellant required to accelerate a given payload increases exponentially with the final velocity of the rocket. The amount of propellant required also varies dramatically with the characteristic exhaust velocity of the propellant (it's "specific impulse").

Some possible rocket propellants

Propellant	Maximum nozzle exhaust velocity meters/second	Minimum % of rocket mass as propellant required to reach exhaust velocity
Hydrogen peroxide, 100%	1,410	99.96%
Black gunpowder	2,000	99.6%
Kerosene + liquid oxygen	3,510	96%
Liquid hydrogen + liquid oxygen	4,460	92%
Liquid hydrogen + liquid fluorine	4,700	91%

I have plotted a graph of the equation above, showing how the amount of propellant increases with the final rocket velocity. Even with an excellent propellant like liquid hydrogen and oxygen, with a high specific impulse, the propellant must comprise more than 92% of the total rocket mass for a rocket to reach escape velocity.

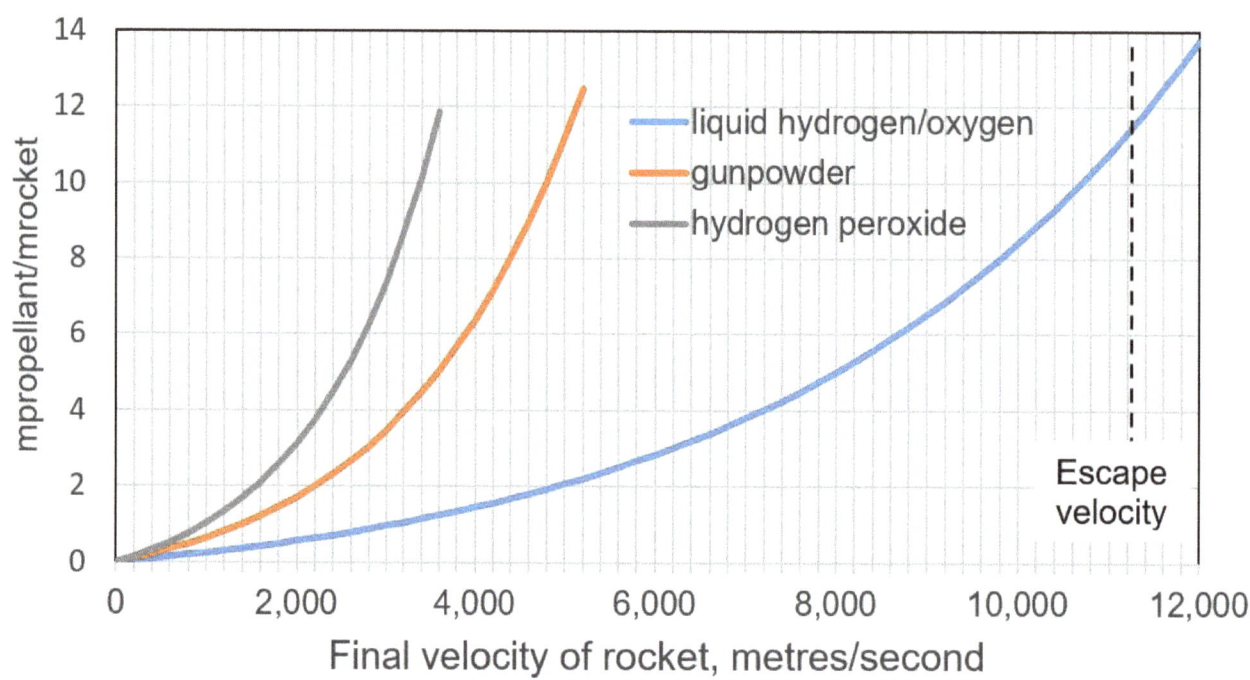

Using fuels with much lower specific impulse, it is virtually impossible to launch a rocket to escape velocity. If we attempted to use black gunpowder as the propellant, it would need to comprise 99.6% of the total mass of the rocket. In other words, the payload, rocket casing, nozzle and other components must comprise less than 0.4% of the total rocket mass. Using current materials and technology, it is inconceivable that a payload could be launched to escape velocity using black powder propellant. If we attempted to use pure hydrogen peroxide as a propellant to reach escape velocity, the hydrogen peroxide would need to comprise 99.96% of the total mass of the rocket! So choosing a propellant system with a very high exhaust velocity is critical for launching rockets into space. This is why rockets used to launch spacecraft generally use liquid propellants with high exhaust velocity. The most commonly used propellant systems are liquid oxygen with liquid hydrogen, and liquid oxygen with kerosene.

The "Rocket equation" has been derived for a single-stage rocket, but the results are applicable to multi-stage rockets as well. What the "Rocket equation" shows is that, to launch a spacecraft to escape velocity, it is essential to use a propellant with very high specific impulse. Even then, the propellant will comprise a very high proportion of the mass of the rocket at launch. The amount of propellant must be far greater than the mass of the spacecraft to be launched into space.

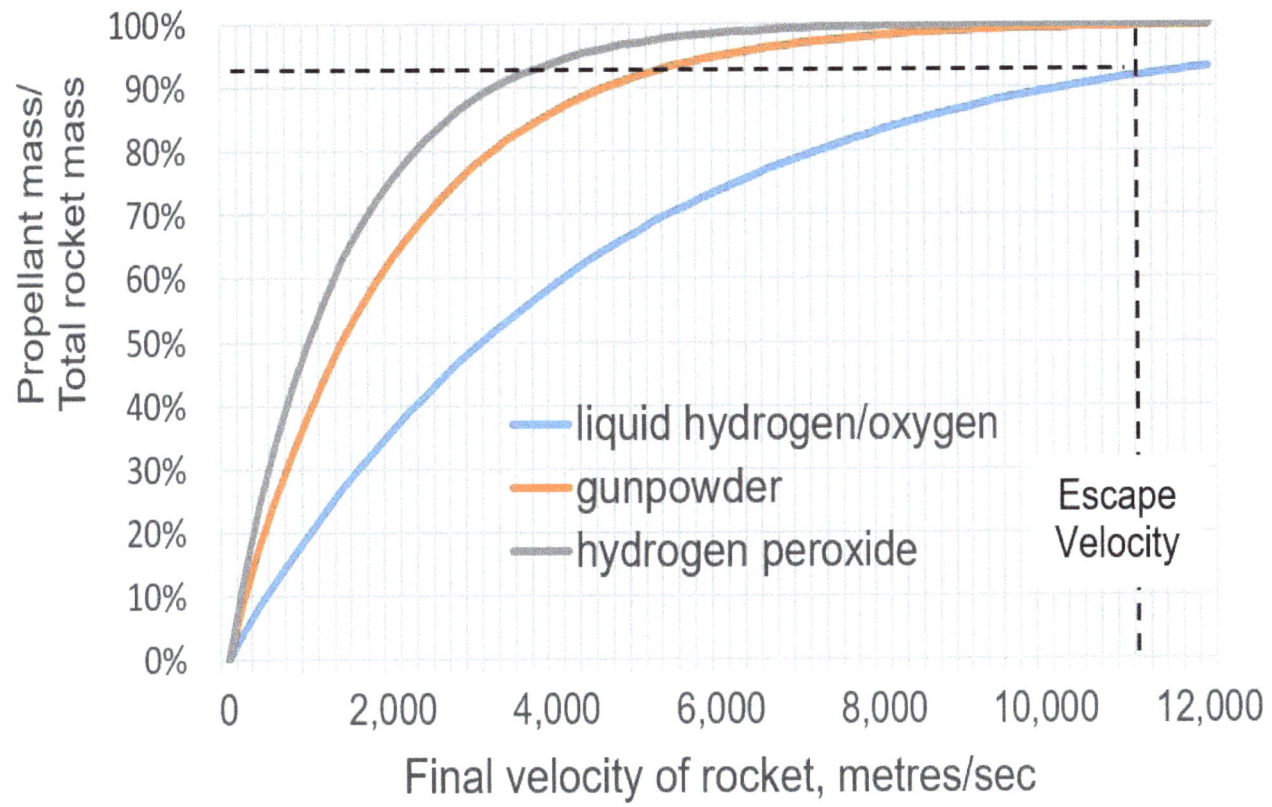

Propellants for contemporary space exploration

One of the great technical achievements of the 20th century was launching of the first astronauts to the moon through the Apollo program operated by the US National Aeronautics and Space Administration (NASA). A number of rockets were developed for the various stages of this program culminating in the development of the Saturn V rocket.

NASA chose to use liquid hydrogen and liquid oxygen as propellant for the second and third stage of the Saturn V rocket, and this propellant combination is still widely used for many space launches. The reaction between hydrogen and oxygen releases a very large amount of energy per unit mass, which heats the product of the reaction, water vapour, to several thousand degrees.

$$H_2 + \tfrac{1}{2} O_2 \rightarrow H_2O$$

If all of the energy released by this reaction is converted to kinetic energy in the rocket nozzle, the water vapour can achieve a maximum velocity of about 4,460 meters/second.

As we have seen, for a rocket using liquid oxygen and hydrogen as propellant, the propellant must comprise a minimum of 92% of the initial mass of the rocket to reach escape velocity. In other words, the rocket itself (fuel tanks, rocket motor, pumps, nozzles, etc *plus* the spacecraft to be launched into deep space) must comprise no more than 8% of the total mass of the rocket at launch.

Above: One of the five rocket engines of the first stage of the Saturn V rocket.

Right: A Saturn V rocket, carrying the first men to the moon on Apollo 11, blasts off on 16 July 1969.

Source: NASA History Office and Kennedy Space Centre.

Considerable engineering skill and technology is required to build a rocket that is light enough to reach escape velocity, yet strong enough to withstand the stresses of blast-off while holding several thousand tonnes of propellant (fuel and oxidizer), pumping hundreds of tonnes of propellant every minute into the rocket engines, which contain gases at high pressure and thousands of degrees. One common approach is to build the rocket in stages, so each stage only has to accelerate the next stage part of the way to escape velocity. Once the fuel in the first stage is exhausted, the entire stage is jettisoned, thereby allowing smaller and less massive fuel tanks, combustion chamber and other components to be used in the next stage.

Only a small fraction of the energy released by the propellant is ultimately imparted as kinetic energy to the spacecraft. In the early stages of the flight, when the rocket is moving relatively slowly, most of the energy liberated by the burning fuel is converted to kinetic energy of the exhaust gases traveling backwards. Then, as the rocket accelerates to hypersonic speed, energy is dissipated in overcoming air resistance as the rocket is driven through the atmosphere. Later in the flight, much of the energy ends up as kinetic energy imparted to the empty rocket stages, which are jettisoned (and eventually fall back to Earth or sometimes, remain in orbit).

The Apollo program to launch US astronauts to the moon (including four successful moon landings, and the near-disastrous Apollo 13 mission) was based on the Saturn V rocket, the largest space rocket built to the current time. Each of the five first-stage rocket engines could lift the weight of 12 railway locomotives.

The Saturn V had a total mass (fuel/oxidizer propellant, rocket engines and payload) at launch of nearly 3,000 tonnes. This comprised:

- 2,700 tonnes of fuel (kerosene in Stage 1; liquid hydrogen in Stages 2 and 3) and oxidizer (liquid oxygen). The fuel and oxidizer (propellant) was 92% of the total mass at launch – very close to the calculated minimum requirement.

- The Saturn V rocket itself (rocket engines, fuel tanks, pumps, etc) had a mass of 187 tonnes, and comprised 6.3% of the launch mass.

- The payload sent to the moon, including the lunar lander and re-entry vehicle had a mass of 47 tonnes - only 1.5% of the launch mass of the fully fuelled rocket). Most of the mass of the payload sent to the moon was fuel and oxidizer for the return journey! (to reach the escape velocity of the moon).

We have seen that the minimum energy required to launch each kilogram of payload into deep space is 63.5 million Joules. But the actual energy requirement is much higher. For the Saturn V rocket, only about 11% of the energy content in the fuel/oxidiser propellant was actually converted to the kinetic energy of the spacecraft sent to the moon.

Bringing rocketry down to Earth

Rockets used to launch payloads into space are huge, complex and use highly-energetic chemical reactions, but this is not necessarily the case for all rockets. Some toy rockets use compressed air to shoot a stream of water downwards through a nozzle. A five-minute video showing a single-stage water (compressed air) rocket can be viewed at:

https://www.youtube.com/watch?v=R625vwA4jpQ&list=PLI1bDyO6U5x4S83dKxMLvGCJSXCiof-71

Small pre-packaged rocket motors containing gunpowder are commercially available for hobbyists who make their own rockets from cardboard and balsa wood (or who buy these as kits). Although these rockets are quite small, the results can be quite spectacular, with the rockets reaching hundreds of meters in height). Some hobbyists fit a simple movie camera to record the flight. A video of such a model rocket flight can be viewed at:

https://www.youtube.com/watch?v=6s7pryfXfHk

If you can get a DVD of the movie "Apollo 13", this is a great movie well worth watching. After reading these notes, you'll have a really good understanding of what's going on, and it will keep you on the edge of your seat. This is a truly heroic adventure story, and it's all true!

6. Tidal forces

During the 1960s, video images of astronauts floating freely within their spacecraft were often shown on television. The terms "weightlessness" and "zero gravity" were widely used to describe the total absence (or, what appeared to be the total absence) of the effect of gravity acting within a spacecraft in orbit. However, it was soon realised that very small gravitational effects were acting on objects floating freely inside an orbiting spacecraft. Objects that were not located at the centre-of-mass of an orbiting spacecraft are subject to a gravity force that is about one-millionth that of normal gravity at the surface of the Earth.

As we have seen, at the centre-of-mass of an orbiting spacecraft, the force of gravity (the attractive force to the centre of the Earth) is exactly equal to the force required to accelerate the spacecraft inwards toward the centre of the Earth and keep it in orbit. However, an object on the side of the spacecraft closer to Earth experiences a slightly stronger gravity force (since the gravitational force varies inversely with the square of the distance to Earth's centre). At the same time, slightly less force is required to accelerate the object inwards to keep it in orbit, since (with its smaller orbital radius and less distance to travel around the orbit) the object must travel at slower velocity to maintain its position within the spacecraft. The overall result is a net force pulling the object towards the Earth.

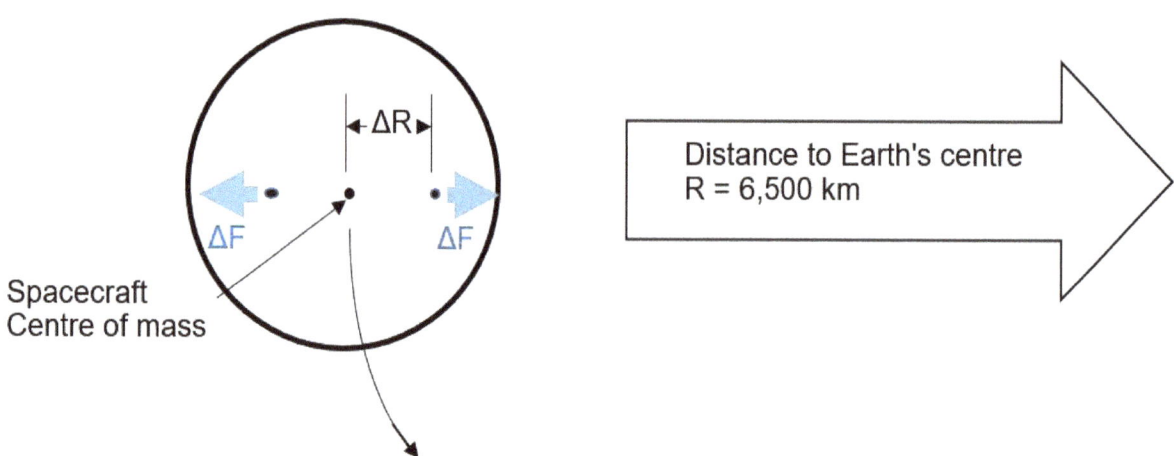

Exactly the opposite situation applies for an object located on the other side of the spacecraft, facing away from Earth. Being slightly further from Earth, this object experiences weaker gravitational attraction to the Earth. At the same time, the object has a slightly larger orbital radius **R** and must travel faster to cover the greater distance along its orbit (**2πR**), so greater force is required to accelerate the object inwards and maintain the object in its orbit. The overall effect is a net force pushing the object away from the centre of the orbiting spacecraft.

Consider a satellite in low-Earth orbit, located about 100-200 kilometres above the Earth's surface (about 6,500 km from the Earth's centre). An object located 2 metres from the centre-of-mass of the satellite (on the Earth-facing side), experiences a net force of about one-millionth of its normal weight on Earth, pulling it towards earth. Thus, the term "microgravity" is appropriate to indicate the scale of such tidal forces. Bear in mind that the tidal force varies directly with the distance from the centre of mass of the spacecraft, so the term "nanogravity" might be more appropriate to describe the tidal forces experienced on the scale of an ant floating within a spacecraft.

In fact, the small tidal force **ΔF** acting on an object located a small distance **ΔR** from the centre-of-mass of a spacecraft can readily be shown (for anyone with a knowledge of 1st year university calculus) to be:

Equation (1) $\quad \Delta F = 3F\left[\dfrac{\Delta R}{R}\right]$

Where   R is the distance to the centre of the Earth
         F is the normal weight of the object on the surface of the Earth

It is clear that the tidal force will be extremely small where the distance from the centre-of-mass (**ΔR**) is about a million times less than the distance to the centre of the Earth (R).

However, tidal forces can actually get quite big where the distance from the centre-of-mass of an object can be thousands of kilometres, and the tidal forces acting on many small pieces of matter add to each other act in a cumulative way. This often applies where a planet is orbiting a star, or a moon is orbiting a planet.

Let's consider the case of the moon orbiting the Earth (or rather, the moon and Earth orbiting each other around their common centre-of-mass). The gravitational attraction of the moon causes water in the Earth's oceans to be pulled into a tidal bulge on the side facing the moon, and another bulge on the side facing away from the moon. This is depicted in the diagram below (not to scale). Even relatively small tidal forces acting on billions of tonnes of water over a distance of thousands of kilometres can cause large effects. On average, this tidal bulge of the oceans is on the scale of 1-2 metres, but varies greatly from location to location, depending upon ocean depth and local topography.

The two tidal bulges move along the surface of the Earth as the moon rotates in relation to the Earth. This *relative* rotation is mainly due to rotation of the Earth on its axis every 24 hours, minus the rotation of the moon around the Earth every 30 days or so.

Another tidal force is caused by the gravitational attraction to the sun, and this adds (or subtracts) from the tidal bulge caused by the moon. The tidal effects of the sun and moon act essentially independently of one another. The tidal effect caused by the moon is about twice as great as that caused by the sun, so let's consider the consequences of the lunar tidal effects a little further.

A considerable amount of work is required to move billions of tonnes of ocean water around the world, overcoming turbulence and friction. The energy source for this work is the rotational energy of the Earth as it spins on its north-south axis. Because of work done by tidal forces, the rotation of the Earth is gradually slowing over a period of millions of years. The same drag force that is slowing the Earth's rotation is increasing the velocity of the moon in its orbit around Earth. This process has been occurring over the billions of years that the moon has been orbiting the Earth.

There is good evidence that the moon was formed by the off-centre impact of a small planet with the Earth about 4.5 billion years ago. This impact is believed responsible for the 23.5° tilt of the Earth's axis (and the rotational plane of the moon) that is responsible for the seasons. At the time that the moon first formed from rocky debris blasted into space by the impact, the Earth would have rotated much faster. The length of a day on Earth would have been much shorter than the current 24 hours. The moon was much closer (and would have appeared much larger than it does today) and rotated around the Earth in a much shorter period of time than it does today. Over the billions of years since, tidal forces slowed the Earth's rotation, and accelerated the moon to an orbit further away from Earth.

We know (from precise measurements of laser beams sent to the moon, and bounced back from reflectors left on the moon's surface by the Apollo astronauts) that the moon is currently receding away from Earth at the rate of 35 mm per year. This is about the same rate as Australia is shifting northwards, and also about the same rate that your fingernails are growing. The Earth's rotation is slowing by about 2 seconds for each 100,000 years, and in about 150 million years, one Earth day will last 25 hours.

Not only is the moon exerting a tidal force on Earth, but the Earth exerts a tidal force on the moon. Of course, there is no ocean of liquid water on the moon, but the tidal forces also cause rock to deform and flow (and indeed, to form a tidal bulge on the side of the moon towards and away from Earth). A considerable amount of work is required to squash, stretch and deform rock, and the energy to do this has come from the rotation of the moon. Over billions of years, these tidal forces have caused the moon to slow its rotation until it exactly matches the period of its rotation around the Earth. The same side of the moon always faces the Earth, and we never see the other side of the moon. From the perspective of an astronaut standing on the surface of the moon, the Earth is in geostationary orbit around the moon! Another way of saying this is that the moon has become "tidally locked" to the Earth.

Tidal forces can be extremely large near massive objects. You may recall that when comet Shoemaker-Levy crashed into Jupiter, it was broken up by tidal forces into a string of objects (resembling pearls on a string), each of which left a visible mark on the surface of Jupiter after impact. The entire event was filmed through telescopes, the first time that such an event has been observed.

Comets are basically collections of rock and ice rubble, loosely held together by the gravitational attraction of its constituents. Consider a piece of rock lying on the surface of Comet Shoemaker-Levy as it hurtled towards Jupiter. At a certain distance from a planet, tidal forces pulling the rock away from the surface (towards or away from Jupiter) exceeded the gravitational attraction of the rock towards the centre of the comet. The particular distance from a planet at which a comet or moon will be pulled apart by tidal forces away is called the "Roche Limit". Using only the equations for gravitational attraction and tidal forces given earlier, the Roche Limit can readily be calculated. It turns out that the Roche Limit depends on the densities of the planet and the impacting comet, and is typically about twice the planet radius (that is, at one planet radius above the planet's surface).

The situation of the Earth and its moon is the opposite of tidal interactions between the planet Mars and its inner-most moon Phobos. The radius of Phobos' orbit is only about three times the radius of Mars, with Phobos completing each orbit around Mars in just 7 hours – less than one-third the time for Mars to rotate on its access. Dissipation of energy by tidal stretching is causing Phobos to lose kinetic energy and slowly spiral **towards** Mars (at the same time, causing Mar's rotational speed to increase). The radius of Phobos' orbit shrinks about 20 millimetres every Earth Year. At that rate, the radius of Phobos' orbit will come within the Roche Limit within 50 million years. Phobos is believed to be made of a loose collection of rock rubble and ice, so that - once inside the Roche Limit - tidal forces will pull some material from Phobos to the surface of Mars and probably break up the remainder into a planetary ring.

Tidal forces are even more spectacular near black holes. A Black hole is an incredibly bizarre remnant of a massive star that has undergone a supernova explosion at the end of its life, with the core of the star collapsing under its own weight. It is believed that the entire mass of the star collapses into a point of infinitesimal size and infinite density, called a "singularity". Nothing – not even light - can escape the gigantic gravitational force near the singularity. However, light can escape if it is emitted at a sufficient distance away from the singularity, outside the "event horizon" of a black hole.

For a black hole with a mass that is ten times the mass of our sun, the event horizon would be something like 60 kilometres in diameter – that is, unbelievably tiny for such an incredibly massive object. Such a black hole would have a mass that is 3 million times greater than the Earth within a volume that is one ten-millionth that of the Earth. An object approaching near to the event horizon would be attracted by a gravitational force about 100 billion times greater than its weight at the surface of the Earth.

Tidal forces near the event horizon of such a black hole would be enormous. Since (from Equation 1) tidal force depends on the gravitational force **F** and the distance **R** from the centre-of-mass, tidal forces at the event horizon would be about **20 trillion times** greater than in low-Earth orbit. Any ill-fated spacecraft, or other object, that approached too close to the event horizon would be accelerated to the speed of light and then disappear forever (never to be seen or heard from again) as it passed through the event horizon. However, before the object reached the event horizon, tidal forces would stretch the object into a long string, even pulling apart the chemical bonds holding the atoms together!. This process has been termed "spaghettification".

Videos

A brief (somewhat frenetic) overview of tidal forces. 2 minutes
  http://www.wimp.com/tidesexplained/

Discussion of tides on Earth and how organisms live in intertidal zone. About 10 minutes
  http://oceanexplorer.noaa.gov/edu/learning/player/lesson10.html

Jupiter's moon Io is subject to huge tidal forces due to its close orbit to this massive planet. Tidal forces stretch and squeeze the rock within Io's crust, causing it to melt into liquid magma. As a result, Io is the most volcanically active body in the solar system. Here is a video of a volcano on Io taken from a spacecraft as it flew past the moon.   17 seconds
  https://www.youtube.com/watch?v=4T37w801UYA

7. The birth, life and death of stars

Within the past 50 years, there has been a revolution in our understanding of the universe. Until about 15 years ago, nobody had any idea whether our star was the only one - among the approximately 100 billion stars in our galaxy – to have planets, or whether most stars have planets orbiting around them. Since then, astronomers have detected about 1,000 exoplanets (planets orbiting other stars), and it is now very clear that many stars have planets.

For those planets that have been discovered around other stars, astronomers can work out their orbital period (how long it takes to orbit their star), their mass and how far these planets are from their star. From this, much can be inferred about conditions on these planets. Some are rocky planets (like Earth), while others are gas giants (like Jupiter and Saturn). Some orbit very close to their star and are scorching hot, while others are freezing cold. Some are in highly eccentric orbits, so that they alternate between scorching hot and freezing cold. A few are believed to be in "the Goldilocks zone" where liquid water (and thus, potentially life as we know it) could exist.

We also have learned that stars vary enormously (sometimes by a factor of *millions*) in lifetime, brightness, density and size. The mass of a star is the most important factor that determines its properties. The second most important factor is the age and history of a star. Some stars have properties that are quite different (some might say "bizarre") compared to our sun, and these are given special names ("brown dwarfs", "red dwarfs", "red giants", "white dwarfs", "neutron stars", "pulsars", "black holes" and "supermassive black holes").

Some stars exist as pairs (or several stars) that orbit around each other. These are called binary stars, and different types of stars can orbit each other. This can lead to strange consequences when one of the orbiting pair is a very dense star (a neutron star or white dwarf, for example). Then, the enormous tidal forces generated by the dense star can gradually suck gas from its orbiting companion, causing the dense star to gradually gain mass. Once its mass exceeds a critical threshold, the denser star literally collapses under its own weight, and explodes into a supernova. On average, there is about one supernova explosion in our galaxy about every hundred years. The last occurred in 1987. Supernova explosions are one of the most spectacular and violent events that occur. Over a few days or weeks, the exploding star produces more light, x-rays and gamma rays than all other stars in the entire galaxy.

A star is born

Our sun is a medium-size, middle-aged, star, halfway through its lifetime of about 10 billion years. When astronomers look across our galaxy with powerful telescopes, they can see regions of space where stars are being born and stars that are dying. While stars continue to be born, the universe seems to have passed its most profligate period of star formation about 8 billion years ago, when stars were being formed at 10 to 20 times the present rate.

Newly-formed stars on the outskirts of the Small Magellanic Clouds in the outer suburbs of our galaxy (Source: Hubble Telescope, http://www.spacetelescope.org/images/heic0702a/)

Stars are born from the gravitational collapse of giant "gas clouds". These gas clouds are thought to comprise about 90% of the known matter in the universe, and fill vast voids within galaxies, and perhaps even within the dark, apparently empty regions between galaxies. Bear in mind that these "gas clouds" are nothing like we think of as "gas" or "clouds". They consist primarily of lone molecules of hydrogen wandering through the vastness of space, with a density of about one molecule per cubic centimetre. This is about 10 million times less dense than the best vacuum that can be produced on Earth. But, on the other hand, these gas clouds are gigantic, so they contain huge numbers of molecules.

The collapse of gas clouds is usually triggered by the shock wave from a supernova explosion, collision between galaxies, or other events whose impact extends across vast regions of space. These shock waves compress the diffuse gas, slightly increasing its density beyond that of the surrounding gas.

If gas were uniformly distributed throughout the galaxy, gas molecules would be attracted equally in all directions by gravitational attraction to the surrounding gas. However, compression of the gas cloud increases its density above the surrounding gas, giving rise to a net gravitational force pulling the gas towards the centre of the gas cloud. When referring to a "gas cloud", we'll only need to consider the region of space where the gas density has been compressed above that of the surrounding gas. Furthermore, we consider the cloud's mass to be the *extra* mass due to the density of the gas cloud being greater than the surrounding gas. The mass within this gas cloud pulls gas molecules towards its centre.

As we will see, such gas clouds are gigantic. A cloud that will ultimately collapse into a star the size of our sun will have an initial radius more than 3,000 times the distance from the sun to the Earth (about the distance that light travels in 3 weeks). During the collapse of the gas cloud, its diameter will reduce more than 3 million times, and its volume by more than 30 billion billion times. This collapse will release an enormous amount of gravitational potential energy. Using a few simple concepts that we have already discussed, we can work out the conditions that are required for the gas cloud to collapse, the amount of energy that is released, and the end result. I'll present a mathematical derivation, which some readers might follow in detail, while others might simply skim through to get the main idea and concentrate on the end result.

Let's assume that the initial gas cloud is spherical in shape, has a uniform gas density ρ, and has a total mass **M**. Let's imagine a small volume of gas with a mass of, say **m** kilograms, located at the edge of the gas cloud at a distance **r** from its centre. We can readily calculate the gravitational potential energy of this gas parcel - that is, the amount of energy that would be released by bringing the gas parcel into the cloud from distant parts of the galaxy or interstellar space. This is exactly the same as the escape energy of the gas parcel – the energy required to pull its mass **m** away from the cloud into distant space, beyond the gravitational force of the gas cloud.

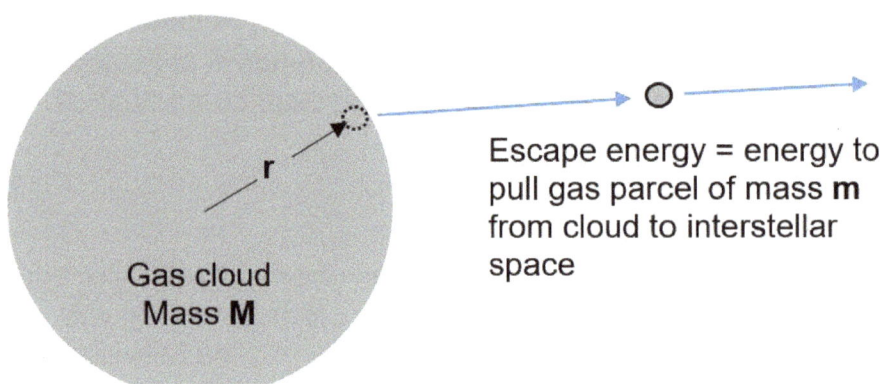

We have previously seen that the escape energy for mass **m** at a distance **r** from the centre of a spherical body of mass **M** is given by:

Equation (1) $$\text{Escape energy} = \frac{GMm}{r}$$

Where G = Universal gravitational constant, 6.67×10^{-11} m³/kg-sec²

To determine the total gravitational potential energy of a gas cloud – the total amount of energy required to disperse **all** the gas into interstellar space – we simply need to "integrate" Equation (1) for every parcel of gas within the cloud. This requires only a knowledge of first year university calculus, and would be a useful exercise for readers with such a mathematical knowledge. The solution is outlined at the end of this chapter (See Note 1).

The result obtained for the total gravitational potential energy of the gas cloud is:

Equation (2) $$\text{Total gravitational potential energy} = \frac{3}{5}\frac{GM^2}{r}$$

It is a simple matter for readers with a basic knowledge of calculus to "differentiate" Equation (2) to determine the amount of energy released when the gas cloud (with a fixed total mass **M**) begins to contract, with its radius reducing by **Δr**.

Equation (3)   Energy released by contraction of gas cloud $= = \dfrac{3}{5}\dfrac{GM^2}{r^2}\Delta r$

From these equations, we can now deduce some amazing insights and conclusions into how our sun was formed. For starters, we will see that a gas cloud can only collapse into a star if the cloud is big enough, and massive enough. We can work out exactly how big, and how massive, the gas cloud must be.

For the gas cloud to contract under its own gravity, the gravitational potential energy that is released when the cloud begins to contract must be enough to compress the gas cloud. After all, the gas cloud is a gas, and it exerts outward pressure.

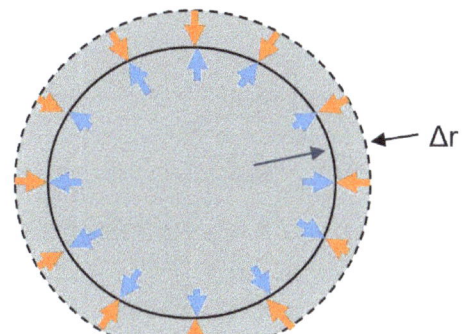

Contraction of a spherical gas cloud (depicted by orange arrows), reducing its radius by Δr, releases gravitational potential energy. For the contraction to occur, the energy released must be greater than the work required to push against the outward pressure of the gas (blue arrows).

We can relate the outward pressure of the gas cloud to its density **ρ**. Molecules within the gas cloud are initially at very low density (in fact, extremely low density) and at very low temperature. Under these conditions, the gas cloud behaves as an "ideal gas", whose behaviour is completely understood and given by the "Ideal Gas Law" (See Note 2). The outward pressure of the spherical gas cloud is given by:

Equation (4)   Outward pressure $= \left[\dfrac{3}{4\pi}\right]\dfrac{MRT}{m_w r^3}$

Where **M** is the total mass of the gas cloud
R is the Universal Gas Constant, 8.3 Joules/mole-degree Kelvin
T is the temperature of the gas (relative to absolute zero)
m_w is the molecular weight of the gas, expressed in kilograms
(0.00175 for gas comprised of 75% hydrogen and 25% helium)
r is the radius of the gas cloud

If the work required to compress the gas cloud is less than the gravitational potential energy that is released, then – and only then – can the gas cloud contract. The condition is given by:

Gravitational potential energy   >   Mechanical work required to
released by contraction   compress the gas

The mechanical work to compress the gas is equal to the gas pressure (given by Equation 4) multiplied by the change in volume of the gas cloud when its radius reduces by **Δr**. The condition for the gas cloud to contract is met when:

$$\frac{3}{5}\frac{GM^2}{r^2}\Delta r \quad > \quad \underbrace{\left[\left(\frac{3}{4\pi}\right)\frac{MRT}{m_w r^3}\right]}_{\text{Gas pressure}} \underbrace{\left[4\pi r^2 \Delta r\right]}_{\text{Volume change}}$$

By simplifying and solving this equation, we find that the gas cloud can only begin to contract if its radius (and therefore, its total mass) exceeds a critical threshold (termed the "Jean's radius").

If the gas cloud is smaller than this critical radius and mass, the cloud will actually diffuse outwards and disperse throughout the galaxy.

The minimum radius for the gas cloud to contract is given by:

$$\text{Minimum radius for gas cloud to contract} = \frac{1}{5}\frac{GMm_w}{RT}$$

So, how big would have been the gas cloud from which our sun formed? If we substitute the mass of our sun (1.9×10^{30} kg) for **M**, and assume that the initial temperature **T** of the gas cloud is 10° above absolute zero, we find that the gas cloud must have had a radius about a million times larger than the current radius of the sun or more than 3,500 times the current distance of the Earth from the sun.

Based on the mass and calculated initial volume of the gas cloud from which our sun originated, we can easily work out the initial density of the cloud, which corresponds to three molecules per cubic centimetre.

Once a gas cloud begins to contract, its temperature rises (as gravitational potential energy is released as heat), but its density increases at an even faster rate. The more the cloud contracts, the greater is the gravitational pull towards the centre of the cloud, until it completely overwhelms the outward pressure of the gas. The contraction accelerates as the cloud decreases in size and increases in density. While contraction of the cloud begins slowly, once it begins, its collapse becomes inevitable.

A disturbance such as a supernova explosion, sends shock waves through the diffuse gas within a galaxy. The gas is compressed within a huge region of space, forming a gas cloud whose density exceeds that of the surrounding gas.

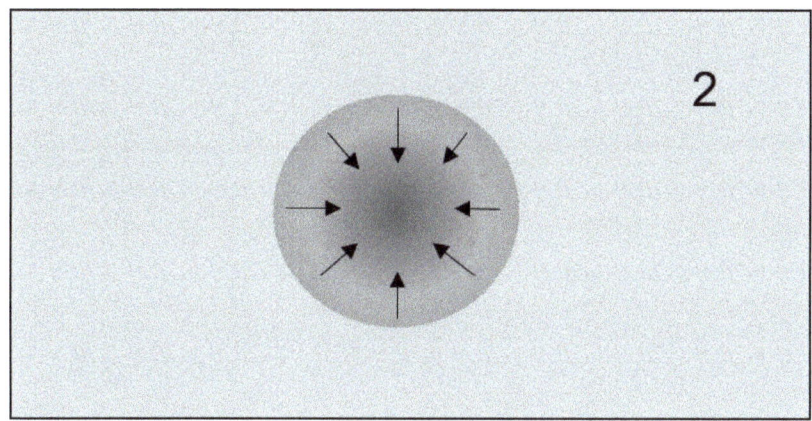

The concentration of mass within the gas cloud exerts gravitational attraction to the centre of the cloud. If the radius (and mass) of the cloud exceed the "Jean's radius", gravitational attraction is strong enough to compress the gas within the cloud, which begins to contract.

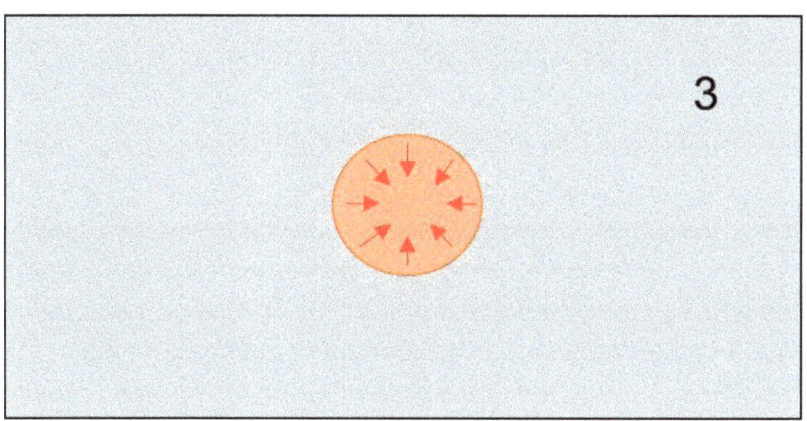

As the cloud contracts, it is heated by the release of gravitational potential energy. However, the increase in temperature is more than offset by the increasing density of the cloud, causing the contraction to accelerate.

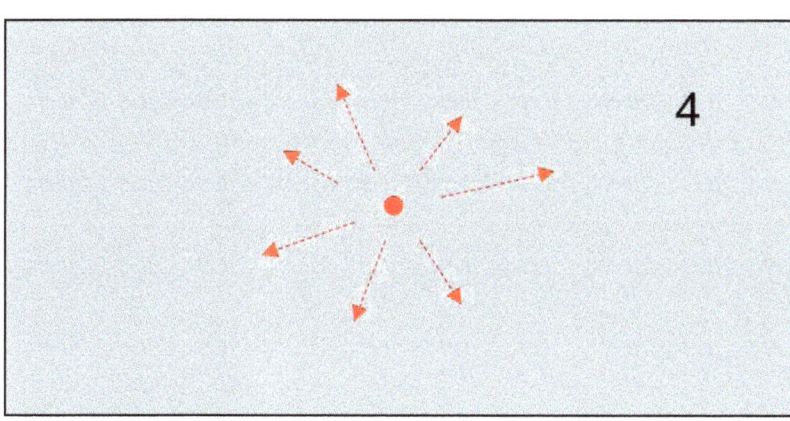

Eventually, when the cloud has contracted to about one-millionth of its original diameter (or one trillion trillionth of its volume), the gas cloud is heated to millions of degrees at its core, hot enough for nuclear fusion to begin. The fusion of hydrogen into helium provides the heat required for the star to maintain its core temperature, remain stable and emit light radiation.

The fate of a star

Once a gas cloud begins to collapse under its own gravity (and its radius **r** shrinks), the gas is heated to higher and higher temperatures by the gravitational potential energy that is released. At temperatures of several thousand degrees, the gas becomes a plasma. The atoms and molecules hit each other with such speed that molecules of hydrogen are broken apart into protons and electrons. The plasma is like a "soup" of protons, electrons and helium nuclei flying around and smashing into each other.

The fate of a collapsing gas cloud is determined by its mass. Let's consider what would happen for gas clouds of various sizes.

If the mass of the proto-star is relatively small – less than about 8% the mass of our sun – the gravitational energy released (according to Equation 1) could heat the proto-star as much as 2 million degrees as it contracted to the same density as our sun. In fact, its actual temperature rise would be somewhat less, since the proto-star begins to lose energy by radiating visible and infrared light from its surface as its temperature rises (and smaller stars have a larger surface area relative to their mass). The temperature at the central core of the proto-star never gets hot enough to initiate nuclear fusion. The proto-star never "ignites". It is destined to be a "brown dwarf", a failed star that glows dimly and gradually fades into darkness.

However, if the mass of the gas cloud exceeds about 8% the mass of our sun, the energy released by its gravitational contraction is sufficient to heat the plasma to tens of millions of degrees. The extremely high temperatures and densities that develop near the central core of the proto-star enable the process of nuclear fusion to begin. Four protons (nuclei of hydrogen atoms) and two electrons fuse into a helium nucleus.

A short (2-1/2 minute) video explaining nuclear fusion can be viewed at:
https://www.youtube.com/watch?v=7E-0j90Cwpk (0 is zero)

This nuclear reaction releases a huge amount of energy – indeed, it is the primary source of energy for stars. The heat released by nuclear fusion keeps the core of the star at such high temperature and pressure as to avoid any further shrinking or collapse of the star. The beginning of nuclear fusion is effectively the moment that a star is born. The huge inwards gravitational force of the star is balanced by the pressure caused by the extremely high velocities of the atomic nuclei and electrons. As long as the fusion process continues, the star remains stable.

Depending on the mass of the star, this process may continue for billions of years, with fusion of hydrogen into helium providing energy to maintain the core of the star at very high temperatures. The outwards pressure of the extremely hot gas balances the inwards force of gravity. Stars that derive their energy by fusing hydrogen into helium are said to be in the "main sequence". Stars spend most of their lives in the main sequence, and these are most of the stars that we see.

A somewhat technical video (9 minutes duration) describing main sequence stars, can be viewed at:
https://www.youtube.com/watch?v=2Lsj5EIuI0o

Eventually stars will fuse all the hydrogen in their core, and "run out of fuel". However, the rate at which stars "burn" hydrogen into helium varies enormously, depending on the mass of the star.

Stars with relatively low mass are just hot enough and dense enough to sustain nuclear fusion at a slow rate. These "red dwarf" stars are relatively small. Their small surface area and low surface temperature (about 3,000°) causes their light emissions to be weak and red in colour. These are among the dimmest stars (and most difficult to see, even with powerful telescopes), but they have very long lifetimes – probably hundreds of billions of years. Red dwarfs are the most common stars in our galaxy although, because they are so dim, none are visible to the naked eye from Earth.

On the other hand, Very massive stars develop extremely high temperatures and pressures (indeed, they need extremely high temperatures and pressures to balance the crushing gravitational force of their huge mass). Such massive stars are extremely bright, and their high surface temperature causes their light emissions to be blue in colour. They "live fast and die young". Massive stars deplete their hydrogen within only 10 or 20 million years.

There is an upper limit to the size of a star, called the "Eddington Limit" (which was originally calculated by a British astronomer named Arthur Stanley Eddington). The larger a star is, the hotter is its surface and the more intense is its emission of radiation. Eddington calculated that if the mass of a star exceeds 108 times the mass of the sun, the radiation near its surface is so intense that "radiation pressure" pushes away the outer layers of the star. Thus, if supermassive stars are formed, they shed their outer layers of gas, until their mass falls within the Eddington Limit.

So, depending on its mass, the time for a star to fuse all the hydrogen within its core can vary from 10 million years (in cosmic terms, the blink of an eye) up to hundreds of billions of years (much longer than the current age of the universe). What happens to a star once it consumes the hydrogen within its core, and the star leaves the main sequence, also differs dramatically – again, depending upon the mass of the star.

The death of stars

Our sun is a mid-size star that is about half-way through its life "on the main sequence" of about 10 billion years. In about five billion years, when the sun has fused all the hydrogen in its core, it will swell up into a Red Giant. It will start to fuse hydrogen and helium in the layers outside the core, forming elements with larger atomic nuclei (particularly carbon, oxygen and nitrogen). As Red Giants expand and shed their outer layers, they release these elements into surrounding space. Red Giants are believed to provide most of these lighter and more abundant elements that are critical to life as we know it [Ref 1].

In becoming a Red Giant, a star increases in volume about a million times, and increases in brightness by 1,000 times. When our sun enters the Red Giant phase of its life, it will swell in size to engulf and incinerate the Earth. Because of their relatively cool surface temperature (about 3,000°), Red Giants emit light mostly in the infrared and red end of the spectrum (and thus, have a red colour).

If we want to see a preview of the fate of our sun in the distant future, we can look around the local neighbourhood within our galaxy and see stars that are currently in the Red Giant phase of their life. One good example that is visible to the naked eye is Betelgeuse (also called "Antares", because its red colour is reminiscent of the planet Mars).

The Red Giant phase lasts a relatively short time (about one-tenth of the star's lifetime on "the main sequence"). Once nuclear fusion has ceased altogether, our sun will shrink down to about one-millionth of its current volume (about the size of the Earth), and live the rest of its life sedately as a White Dwarf. The relatively tiny size and surface area of White Dwarfs greatly reduces energy loss through radiation, so they can retain their heat for tens of billions of years before gradually fading (eventually perhaps becoming a "Black dwarf"). White dwarfs are very faint stars, and difficult to see.

That's the ultimate fate of our sun, and for stars of similar mass. The fate of other stars can differ dramatically, depending on their mass.

Very small stars, with a mass less than 40% of our sun, live sedate lives. With lower temperatures and pressures in their core, they fuse hydrogen to helium at a very low rate (they are unable to fuse helium or produce heavier elements). These stars, called "Red dwarfs", have low surface temperatures and are relatively faint. Because of their frugal lifestyle, Red dwarfs have very long lifetimes, lasting **_hundreds of billions_** of years.

On the other hand, stars that are significantly more massive than our sun live frenetic lives, and undergo a spectacular death.

Very massive stars (with up to a hundred times the mass of the sun), burn up their hydrogen within a few million years. Then, as their core begins to collapse under the enormous weight of the star, the temperature and pressure rise until helium can fuse into carbon, oxygen and heavier elements. As larger and larger atomic nuclei are produced, the nuclear fusion reactions produce less and less additional energy. Finally, once chemical nuclei within the core have fused into iron, no more energy can be produced, and the pressure of the gas cannot support the crushing inwards gravitational force.

At this point, the core collapses catastrophically, and the shock wave blasts outer layers of the star outwards at very high speed (about one-tenth the speed of light). The resulting supernova explosion is often accompanied by an incredibly powerful, but very brief, burst of gamma ray radiation. For a period of a few minutes, the "Gamma Ray Burst" can have an intensity one-hundred times greater than all radiation produced by the entire galaxy.

Supernova explosions release an extremely intense stream of neutrons, which are absorbed by the outwards-flying debris. These high-energy neutrons strike atomic nuclei of elements formed in the final stages of the star's life. and in this way, form atomic nuclei even larger than iron. The energy required to form such large, energetically unstable nuclei is provided by kinetic energy imparted to the neutrons during the supernova explosion. In this way, supernova explosions seed the universe with heavy elements. In fact, all elements heavier than iron are thought to have been created and spewed outwards into the universe during the final violent moments of supernova explosions.

Short-lived, unstable radioactive isotopes ejected by the explosion undergo radioactive decay over the days and weeks following a supernova. Energy released by this radioactive decay creates an extremely bright visible glow, giving off more light than the rest of the galaxy combined.

Kerpow!
A supernova explosion

The Crab Nebula is a remnant of a supernova explosion that was observed and recorded by Chinese astronomers on 4th July 1054 (references to it are also found in Japanese and Arab documents). It is located in the large Magellanic Cloud, on the outskirts of the Milky Way galaxy.

The outer layers of the exploding star formed the nebula of gas and dust seen in this photograph, while its core collapsed into a pulsar, a neutron star spinning about 30 times per second. The nebula and the pulsar are the most studied astronomical objects outside the Solar System. It is one of the few supernovae where the date of the explosion is precisely known.

Source:
 NASA (http://www.nasa.gov/images/content/430450main_image_1604_946-710.jpg)

Other heavy nuclei formed in the explosion undergo radioactive decay over a very long period of time – over billions of years in the case of uranium. These long-lived, unstable elements are strewn throughout the surrounding galaxy by supernova explosions. Uranium and other heavy elements are eventually incorporated into dust particles, which then collect into planets that form around new stars. Being so heavy, these nuclei sink deep into the molten crust and core of newborn planets. The gradual radioactive decay of these unstable elements keeps the core of the planet hot and molten. In the case of Earth, heat released by radioactive decay of heavy elements creates convection currents of molten rock that is responsible for volcanoes, earthquakes and the motion of tectonic plates across the surface of the planet. The heat from radioactive decay also keeps Earth's iron core in a molten state, and thereby is responsible for the magnetic field that protects Earth from a "solar wind" of high-energy charged particles emitted by the sun. Consequently, all life on Earth owes its existence to the formation of heavy elements formed during final seconds of ancient stars which ended their lives in supernova explosions billions of years ago.

For stars that are slightly larger than our sun, with a mass between about 1.5 and 3 times that of sun, the supernova explosion at the end of their lives leaves a remnant of a "neutron star". As the star collapses under its own gravity, protons and electrons are crushed together into neutrons. Effectively, the core of the star collapses into a single giant atomic nucleus, composed entirely of neutrons. A neutron star has a radius of only about 10 kilometres, and its density is about 1,000 ***trillion*** times that of the sun. Some neutron stars have an extremely powerful magnetic field that rotates (typically about once per second), sending out powerful pulses of radiation at precise intervals. These are termed "pulsars". Some pulsars are known to have orbiting planets.

Other pulsars orbit another star, in a binary pair. With such high density and tiny radius, the gravity of a neutron star creates enormously powerful tidal forces which may pull gas from its companion star. The neutron star gradually gains mass until it exceeds another critical threshold. The gravitational forces become so crushing, that even a neutron star collapses under its own weight. It undergoes another supernova explosion, forming a "black hole".

The gravitational forces within a black hole are so gigantic that the escape velocity exceeds the speed of light, and not even light cannot escape. The conditions within a black hole are so extreme that the known laws of physics may break down. The best guess is that all the mass shrinks to an infinitely small point called a "singularity", and the black hole is bounded within an "event horizon". Anything falling within the "event horizon" can never escape. Nor could we know what happens inside the event horizon, because no signal sent from inside the event horizon would be able to escape to the outside universe.

Black holes sound pretty scary. They create huge gravitational attraction in their immediate vicinity, which pulls in any mass that happens to wander nearby. And, if that mass is pulled within the event horizon, it will never escape. In this way, "black holes can grow bigger by consuming gas, stars or anything else that wanders into their path.

Astronomers have detected a supermassive black hole, with a mass several million times that of the sun, at the centre of our Milky Way galaxy. Astronomers believe that every spiral galaxy has a supermassive black hole at its centre, and the mass of this supermassive black hole seems to comprise about one-thousandth of the mass of the central bulge of the spiral galaxy. Why supermassive black holes should comprise a nearly constant proportion of a galaxy's mass is not understood, but astronomers suspect that supermassive black holes may have played a key role in the formation of spiral galaxies.

If all this sounds alarming, you might be wondering, "what keeps black holes from swallowing the entire galaxy?"

For a start, black holes are very small. The radius of a "black holes" event horizon is proportional to its mass, so if a black hole had the same mass as the Earth, its radius would be only 9 mm!

Black holes are formed directly when stars that are more than three times the mass of our sun undergo a supernova explosion at the end of their lives. The "event horizon" for the black hole formed in this way would have a radius of about 8 kilometres. This is pretty tiny in cosmic terms, and our galaxy is pretty gigantic. The chance of another star making a direct impact with a black hole is pretty remote.

Furthermore, black holes are "messy eaters". As one astronomer noted[2], rather than being like galactic vacuum cleaners, black holes are more like cosmic leaf blowers. Gas falling towards a black hole often ends up spinning in a close-orbit (called an "accretion disk"), is accelerated to near-light speeds, is heated to millions of degrees by frictional drag with the surrounding gas, and produces intense radiation of x-rays and gamma rays. This high-energy radiation carries momentum away from the black hole, which pushes away other infalling mass. This "radiation pressure" limits how fast a black hole can grow. Physicists calculate that a black hole would need 50 million years to double its mass, even if it sucked in material continually at the maximum rate.

Notes & Mathematical derivations

1. The escape energy is the same for any gas parcel with mass **m** at the same radius **r**. If we consider all the gas within a hollow spherical shell with thickness **Δr** at radius **r**, its mass **m** is given by **ρ(4πr²)Δr**, and its escape energy is given by substituting this expression for mass **m** in Equation (1). To determine the total gravitational potential energy of the entire gas cloud, we add the escape energies for all hollow shells of thickness **Δr** from **r** = 0 to **r** = r. The total mass **M** within radius **r** is the gas density **ρ** times the volume within radius **r** (4/3πr³). In the limit as **Δr** approaches zero, the sum is given by the integral:

$$\text{Total gravitational potential energy} = \int_{r=0}^{r=r} \frac{G(\rho \frac{4}{3}\pi r^3)(\rho 4\pi r^2)}{r} dr$$

This integral can readily be determined, giving the gravitational potential energy in terms of the gas density **ρ**. The density of the gas cloud is given by its total mass **M** divided by its volume, ρ = M/(4/3 πr³). Substituting this expression for the gas density gives Equation (2).

---

2. The Ideal Gas Law relates the pressure **P** of a gas to the number of moles of gas, the gas temperature and its volume. The pressure of an ideal gas is related to its density by:

   Pressure P = ρRT/m_w Where **R** is the Universal Gas Constant, 8.3 Joules/mole-degree
   m_w is the molecular weight of the gas, expressed in kilograms
   (0.00175 for gas comprised of 75% hydrogen and 25% helium)

   For a spherical gas cloud of radius r, its volume is 4/3 πr³, and its density is M / (4/3 πr³)

References

(1) Donald Goldsmith, The far, far future of stars. Secrets of the Universe: Past, present and future. Scientific American, Fall 2014.

(2) Jenny E. Green, Goldilocks Black Holes, ibid

8. The essence of atoms

Major discoveries in the late nineteen and early twentieth century changed our understanding of atoms and their chemical interactions. During this time, three basic components of atoms were identified – protons, neutrons and electrons. Protons and neutrons form the central nucleus of the atom, and comprise nearly all of its mass. Protons have a positive charge, while neutrons (as their name suggests) are electrically neutral. Electrons, which are negatively charged (and thus, attracted to the positively-charged nucleus) move in orbitals around the nucleus.

Compared to the overall size of the atom (that is, the size of the electron orbitals), the nucleus is tiny, comprising about one quadrillionth of its volume. If we imagine an atom as the size of a two-story house, its nucleus would be about 0.1 mm in diameter (about the size of the full stop at the end of this sentence). Thus, in any drawing depicting an atom, the nucleus is invariably shown grossly exaggerated in size (otherwise, anyone looking at the drawing would not see it!).

Despite being so small, the nucleus contains nearly all (about 99.9%) of the mass of an atom. Protons and neutrons have masses that are nearly equal, and the mass of a proton or neutron is called 1 AMU (Atomic Mass Unit). Electrons have a mass that is about $1/2,000^{th}$ that of a proton or neutron.

At first, it was thought that electrons move in circular orbits around the nucleus, exactly analogous to how the Earth and planets move around the sun (with the electrical attraction of the electron to the positively-charged nucleus replacing the gravitational attraction of the Earth to the sun). However, subsequent experiments showed that this concept of the atom was fundamentally flawed. At the scale of an atom, new laws of physics apply. In fact, these laws, which are called "quantum mechanics", apply at all scales, but quantum effects are absolutely insignificant and immeasurable at the normal scale at which we are familiar.

Many readers will recall that, up until about 10 or 15 years ago, the screens of television sets and computers were "cathode ray tubes", which focused a narrow beam of electrons onto a phosphorescent screen. This electron beam was then scanned back-and-forth across the screen to form a picture. In this situation, the electrons are free to move within the cathode ray tube which, as far as the electron was concerned, was a journey through free space (the beam was nearly one millimetre in diameter, and travelled for perhaps 20 centimeters – gigantic compared to the size of an atom). In this situation, the electron acts like a "normal" particle and obeys "classical physics" – physics until the discovery of general relativity and quantum mechanics.

However, within an atom, the electron is confined within a tiny space (within one ten millionth of one millimetre of the nucleus). In this situation, the electron acts more like a wave than a particle – in ways that we would consider quite bizarre. In addition to the "normal" laws of classical physics, the behaviour of the electron is governed by Heisenberg's Uncertainty principle. This states that the position and momentum (velocity) of the electron cannot both be known with certainty. We could design an experiment to know exactly where the electron is located at any point in time, but then we could not know the velocity of the electron. Alternatively, we could design an experiment that tells us exactly how fast the electron is moving at a particular point in time, but then we could not determine where the electron is with any degree of certainty.

So, what does the Heisenberg Uncertainty Principle mean for an electron confined within an atom? Classical physics tells us that the electron is attracted to the nucleus, and "wants" to be near the nucleus (in a lower state of potential energy). But if the electron stays very near the nucleus, its position would be known with a high degree of certainty and it velocity could be very large (and thus, the electron could readily overcome the attraction of the nucleus and move further away). So, the electron "wants" to be near the nucleus, but doesn't "want" to be confined within any small region of space.

To balance these opposing effects, the electron resides within an atomic orbital whose size and shape achieves the best compromise for the electron to be in the region of lowest potential energy, without being confined within too small a region of space.

As a result of the Heisenberg Uncertainty Principle, we cannot say that an electron within an atom is moving in a simple circular orbit around the nucleus. In fact, we cannot say exactly where the electron will be within the atom. But we can determine a probability distribution for the electron – that is – the probability that the electron will be located at each position in space. The electron will be confined within an "orbital", which we can imagine being like a cloud. Darker areas of the cloud, near the nucleus, have highest probability of finding the electron, and paler parts of the cloud (further from the nucleus) have lower probability of finding the electron.

Thus, when we say that a hydrogen atom has a radius of about 0.5×10^{-10} metres, this does not mean that the electron orbital has a sharp boundary at this distance from the nucleus. The orbital extends forever, but beyond this radius the probability of finding the electron rapidly becomes insignificant. Even beyond a distance of 1.0×10^{-10} metres, the probability of finding the electron becomes very small.

An electron orbital as it might look if the shade of colour indicated the probability of finding the electron.

So, let's consider the hydrogen atom, which is the simplest type of atom. Hydrogen is, by far, the most abundant element in the universe, comprising about three-quarters of all the known matter in the universe (in recent years, cosmologists have discovered that most of the mass and energy in the universe are "dark matter" and "dark energy", and we have no idea what these are!!!). Hydrogen comprises most of the mass in stars, and is believed to have formed just after the universe was created in "the Big Bang" about 13.8 billion years ago.

A hydrogen atom contains one proton in its nucleus. "Normal" hydrogen contains no neutrons. However, about 0.15% of hydrogen atoms contain one neutron in their nucleus, and this is called deuterium. Deuterium has almost exactly twice the mass of "normal" hydrogen, but its chemical properties are virtually identical to those of normal hydrogen. The electron is simply attracted to the positive charge of the proton in the nucleus, and doesn't "care" if a neutron is present. Deuterium is called an "isotope" of hydrogen. Isotopes of an element have the same number of protons and electrons – and thus, have nearly identical properties, but differ in the number of neutrons (and thus, differ in mass).

By the way, there is another isotope of hydrogen, called tritium, which has two neutrons in the nucleus. However, tritium is unstable, and undergoes radioactive decay within a few years (one of the neutrons forms a proton, which remains in the nucleus, and ejects a high-energy electron). Consequently, tritium may be formed in nuclear reactors, but is not found in nature. If any tritium had been present in the dust and gas from which the Earth formed, it would have decayed long, long ago.

One of the bizarre features of quantum mechanics is that only two electrons can occupy any orbital. This is true even in the "macro world" that we humans inhabit, but here, the limit of two electrons per orbital is of no consequence. For electrons moving in free space (say, the electron beam within a television "cathode ray tube"), the orbitals are so numerous, and so closely spaced in energy, that the two-electron limitation has no observable effects or significance. But, the smaller the region of space to which electrons are confined, the greater is the energy difference between orbitals.

For a hydrogen atom, the electron orbitals are clustered into "shells". The first shell contains only one orbital which can hold two electrons. A neutral hydrogen atom contains only one electron, so its outer shell is half-full.

As it turns out, the electron orbitals of hydrogen (and other elements) are most stable (have the lowest energy) when they form a complete outer shell. In a hydrogen atom, one orbital comprises its outer shell, and it can accommodate two electrons. Thus, a single hydrogen atom will be very "unhappy" (have high energy) because its outer shell is only half full. It will tend to undergo any chemical reaction that will complete its outer shell (or make the atom "feel" like it has a full outer shell). To complete its outer shell, the hydrogen atom has three options:

- It can **lose an electron** to another atom which wants to gain an electron (non-metals like oxygen or fluorine).

- It can **gain an electron** from another atom which wants to lose an electron (a metal like sodium or potassium), or

- It can **share an electron** with another atom of hydrogen, or another atom with a partly-full outer shell and a propensity to share electrons (like carbon or nitrogen).

If no other elements are present, a hydrogen atom has only one option to complete its outer shell: it can share its electron with another hydrogen atom. This sharing of electrons between two hydrogen nuclei creates a chemical bond, binding the two atoms within one molecule.

To understand why this chemical bonding occurs (that is, why it offers a state of lower energy), imagine the union of two hydrogen atoms as they are brought together. The atomic orbitals on the two hydrogen atoms overlap and merge into a single molecular orbital. From the perspective of each electron, this merger offers an energy-lowering payoff, and an energy-raising cost.

- The electrons are attracted to the lower potential energy region between the two nuclei, where each electron is attracted to both nuclei. That's a very energy-favourable place for the electrons to be.

- On the other hand, each electron is repelled by the other electron, and that raises its energy.

However, compared with the nuclei, the electrons are very nimble. They have very low mass and high velocity, and they can "dance" around each other, moving to avoid the region of space occupied by the other electron at that instant. The electrons can "take turns" to occupy the low energy region between the two nuclei, while minimising their mutual repulsion. Thus, the combined hydrogen molecule has lower energy than two separate hydrogen atoms. The reaction occurs spontaneously when hydrogen atoms come into contact, and the energy difference is released as heat.

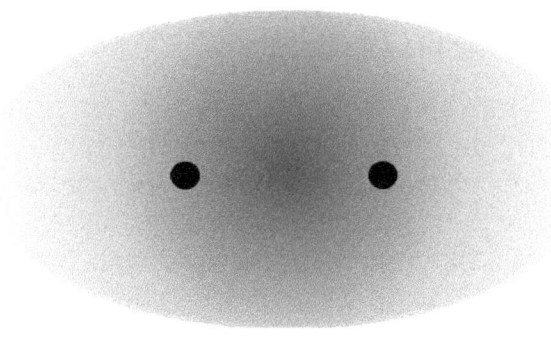

The two electrons then occupy a single "molecular orbit" that binds the atoms into a single molecule.

Thus, if we have a tank of hydrogen gas, it would contain virtually no lone hydrogen atoms. As soon as a hydrogen atom struck another hydrogen atom, the atoms would chemically react to form hydrogen molecules (each molecule containing two hydrogen atoms and represented as H_2). This sharing of outer-shell electrons offers a lower state of energy, as each atom is closer to having a full outer shell, but this does not mean that the hydrogen molecules will not react further.

For example, if a hydrogen molecule encountered a fluorine or oxygen molecule, this would offer a "better deal" (lower energy) in providing each atom with a complete outer shell. Why is this? Well, an oxygen atom has eight electrons. Two electrons fill its first (inner) shell, leaving six electrons in its outer shell. For reasons that are beyond the scope of this book, but of fundamental importance, the second shell comprises four atomic orbitals – and thus, is capable of holding eight electrons. So, oxygen atoms need to gain two electrons to complete their outer shell, and are desperately "looking" for an atom that is not as covetous of its electrons. If they cannot find another such atom, the best that an oxygen atom can do is to share two of their outer shell electrons with another oxygen atom. But, if an oxygen atom (or O_2 molecule) happens to wander into some hydrogen atoms, this offers a great opportunity to go to a lower energy state.

If we were to mix hydrogen gas (containing H_2 molecules) with oxygen gas (containing O_2 molecules), we might expect a reaction to instantly occur. But we would be disappointed.

Initially **nothing** happens when we mix hydrogen and oxygen at room temperature! Before a chemical bond can be formed between the hydrogen and oxygen atoms, energy must be provided to break or weaken the existing hydrogen-hydrogen and oxygen-oxygen bonds. This "activation energy" can be provided by high temperature, a spark, or high-energy ultraviolet radiation. Should we provide a small spark or other energy source to initiate the reaction, the reaction will begin to occur. As soon as this happens, the reaction between hydrogen and oxygen releases substantial amounts of heat, raising the temperature of the gas and providing the ignition energy for more hydrogen and oxygen to react. The result is an explosion (a chemical reaction that occurs faster and faster, as the heat initially released by the reaction causes remaining gases to react).

The tendency for an atom to chemically react with other atoms, with each atom "trying" to achieve a complete outer shell (or the "best deal", providing the most complete outer shell that can be achieved in the circumstances) is not limited to hydrogen. It applies to all atoms. This tendency drives nearly all chemical reactions. It explains which elements or molecules will react, and which molecules they will form. It explains much of the chemistry that we know.

After hydrogen, the next simplest element is helium, which has two protons in the nucleus, (as well as two neutrons, generally). To be electrically neutral, the helium atom must have two electrons, which fill the outer shell. Consequently, helium has "nothing to gain" by reacting with another atom – and in fact – helium does not chemically react with any other element or compound. It is completely chemically inert. Helium is the second most abundant element in the universe, accounting for about one-quarter of all atoms. Like hydrogen, most helium was formed just after the universe exploded into existence after the Big Bang.

All the remaining naturally-occurring elements, up to uranium with 92 protons in its nucleus, comprise a small fraction of the total matter in the universe. However, these elements constitute nearly all the mass of rocky planets like Earth. Without these heavier elements, chemistry would be a very simple and boring subject, which wouldn't really matter, as rocks, trees and life (as we know it) could not exist. Elements other than hydrogen and helium did not exist in the very early universe. They were formed by nuclear fusion within giant stars that were the first to form, and were then spewed throughout the surrounding galaxy when these ancient stars exploded in supernova explosions. These elements formed dust rings around new stars that formed, and eventually collected into rocky planets like Mercury, Venus, the Earth and Mars.

Elements following hydrogen and helium in the periodic table – those having more than two protons - play a fundamental role in our world and in all living things. These elements, forming the second and third row of the periodic table, can hold eight electrons in their outer shell. Here is a simplified periodic table, showing the first three rows:

H							He
Li	Be	B	C	N	O	F	Ne
Na	Mg	Al	Si	P	S	Cl	Ar

The position of these elements in the periodic table shows how many electrons are needed to complete their outer shell. For example, oxygen (chemical symbol O) is in column six of the periodic table. It has six electrons in its outer shell. Consequently, an oxygen atom has three conceivable options to undergo a chemical reaction to achieve a lower state of energy:

1. It can gain two electrons by reacting with any element which has a strong tendency to lose electrons (that is, an element with one or two electrons in the outer shell).

2. It can lose six electrons. This does not occur, as the loss of so many electrons is too difficult (requires too much energy).

3. It can share two electrons with atoms with a similar tendency to gain or lose electrons.

It is important to realise that Options 1 (gaining two electrons) and Option 3 (sharing two electrons) are not necessarily mutually exclusive. These options represent the ends of a continuum. Let me explain. Consider the reaction between hydrogen and oxygen. Each atom can best achieve a complete outer shell if two hydrogen atoms share their electron with one oxygen atom (forming the molecule H_2O). However, the sharing of electrons is not equal. Oxygen has a stronger affinity to gain electrons than hydrogen, and so electrons in the hydrogen-oxygen molecular orbitals are more likely to be near the oxygen nucleus than the hydrogen nucleus. If we could see the molecular orbitals within a water molecule, the probability of finding the bonding electrons near the oxygen nucleus is greater than the probability of finding the electrons near the hydrogen nucleus.

We can imagine the bonding electrons of a water molecule as being within a cloud, with darker shading in areas with the highest probability of finding an electron. An electron spends more time near the oxygen nucleus than near a hydrogen nucleus. While the molecule is electrically neutral overall - with the negative charge of all electrons exactly equal to the positive charge of all the protons – the region around each hydrogen atom has a net positive charge.

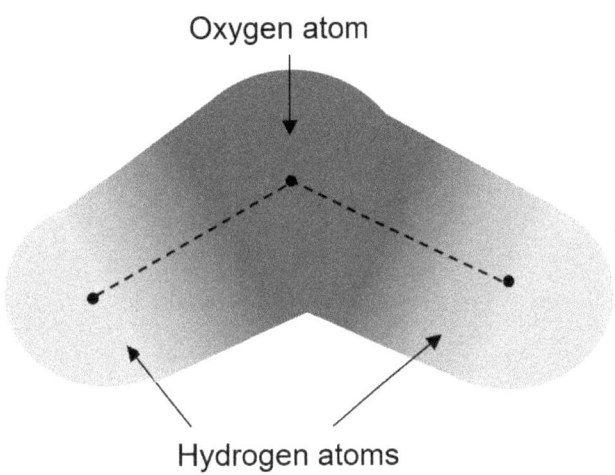

This occurs because the positive charge of the proton in each hydrogen nucleus is not fully offset (or shielded) by the electron charge around it. Similarly, the region around the oxygen nucleus has a net negative charge.

Because of the separation of charge, we say that the water molecule is "polar covalent".

By the way, you might be wondering why I have drawn the water molecule with a "bond angle" (between the central axes of the two hydrogen-oxygen bonds) at about 109°. The electron pairs in the two hydrogen-oxygen bonds are repelled by the two other (non-bonding) electrons pairs in the outer shell of the oxygen atom. This bond angle allows the outer shell electrons to be as far away from each other as possible (along the axes of a tetrahedron).

But let's return to the nature of the chemical bonds holding together a chemical compound.

Basically, we can distinguish three types of chemical compounds:

- Ionic compounds – in which electrons are completely (or nearly completely) transferred from one atom (or group of atoms) to another. An example is common table salt (sodium chloride, NaCl). When these compounds dissolve in water, they form positive and negative ions (atoms, or groups of atoms that have lost or gained an electron). Consequently, solutions of such salts conduct electricity.

- Covalent compounds – in which electrons are equally shared between atoms. Examples are molecules of hydrogen (H_2), oxygen (O_2), fluorine (F_2) and nitrogen (N_2).

- Polar covalent compounds – in which electrons are shared between atoms, but the sharing is not equal, leaving each atom with a partial positive or negative charge. Nearly all organic compounds (those making up our bodies and all living things) have this type of bonding.

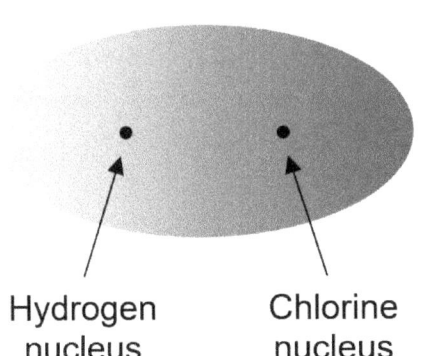

Another example of a polar covalent molecule is hydrogen chloride (HCl), which is held together by two electrons sharing a molecular orbital extending across both atoms. The two bonding electrons have a higher probability of being near the chlorine nucleus, giving this end of the molecule a net negative charge, while the other end has a net positive charge.

Finally, it is possible to mix-and-match different types of chemical bonding. For example, in the compound sodium acetate, an electron is completely transferred from the sodium atom to the remaining complex of atoms ($C_2H_3O_2$), which is held together by polar covalent bonds. When dissolved in water, sodium acetate forms positively-charge sodium ions (Na^+) and negatively-charged acetate ions ($C_2H_3O_2^-$).

Most covalent bonds involve the sharing of outer-shell electrons between two atoms. But covalent bonding holding together solid metals, called (not surprisingly) a "metallic bond", involves billions or trillions of atoms. This type of bond holds together a piece of sodium, aluminium or iron, and is responsible for the strength of metal structures.

It is instructive to consider the formation of metallic bonds if you want to get an intuitive "feel" for the quantum world. I didn't learn about metallic bonds when I studied chemistry in high school or university, however, there is a related application of quantum mechanics covered by most physical chemistry courses – the situation of "a particle in a box". This considers the very simple hypothetical situation of an electron (or any particle) that is confined within a "box" of fixed dimensions. The electron is completely free to wander inside the box, but faces an impenetrable barrier at the wall of the box. This is, very roughly, the situation of an outer shell electron in a metal atom. Within a single metal atom, an outer shell electron can move relatively unhindered within its atomic orbital, but is prevented from penetrating too far into the surrounding space by attraction to its atomic nucleus. A similar situation applies within the crystal structure formed by millions of metal atoms. Here outer shell electrons can wanter relatively freely throughout the **entire piece of metal,** which is effectively "the box" containing the electrons.

The situation of a "particle in a box" is the only problem in quantum mechanics which has a simple solution which can readily be derived by anyone who has completed an introductory calculus course. So, if you would like to bear with me for the next page, we can apply the solution to the "particle in a box problem" to consider the formation of metallic bonds from the perspective of an outer shell electron.

A metallic bond involves outer-shell electrons in metal atoms being shared among many, many (often trillions) of other identical metal atoms. These electrons are delocalised that is, they can freely migrate through the structure of the metal (perhaps metres in length). The presence of delocalised, free electrons is responsible for the unique common properties of metals: they conduct electricity; they have a shiny "specular" surface that reflects light; and they are ductile and malleable (tending to stretch or bend, rather than crack or shatter, when bent).

So, why do the outer-shell electrons in metal atoms choose to wander away from their original atoms and exchange places with neighbouring electrons? Why does this nomadic life offer a state of lower energy?

In a lone metal atom, outer shell electrons are attracted towards the positive charge of the atomic nucleus. Being close to the atomic nucleus is attractive for the electron, but this confines the electron with a tiny region of space occupied by the atomic orbital. The electron is effectively trapped within a "box" of atomic dimensions. The lowest energy orbital that the electron can occupy (its "ground state") is determined by the size of the box in which it is imprisoned.

If metal atoms are brought sufficiently close together, the electron "feels" the attraction of surrounding atomic nuclei, and the energy barrier for the electron to wander from one atom to the next is reduced. The electron can wander through this barrier, and is effectively confined within a much bigger box.

As we allow electrons to move freely from atom to atom, not only does the box get bigger, but we have more and more outer shell electrons sharing the bigger box. But remember that only two electrons can occupy any orbital. As we put more and more metal atoms (and their outer shell electrons) together, the additional electrons must go into higher and higher energy orbitals. Of course, electrons don't "want" to go into orbitals of higher energy. But, at the same time, the size of the box to which all electrons are confined gets bigger and bigger, and this lowers the energy of *all* the orbitals. So, which effect wins out?

In a metal, where there are only one, two or three outer shell electrons, the effect of delocalising electrons within a bigger box wins out. The total energy of electrons in delocalised molecular orbitals is less than within atomic orbitals of separate metal atoms.

Imagine that we have a lump of metal, containing say a few billion metal atoms, and we bring one additional metal atom to the surface. The outer shell electrons of this atom must go into a high energy molecular orbital (since the lower energy orbitals are all filled), but – by slightly expanding the size of the metal piece (the "box"), the energy is reduced for *all* the other billions of delocalised electrons within the metal. The total energy for all electrons is reduced by the addition of each metal atom to the metal structure.

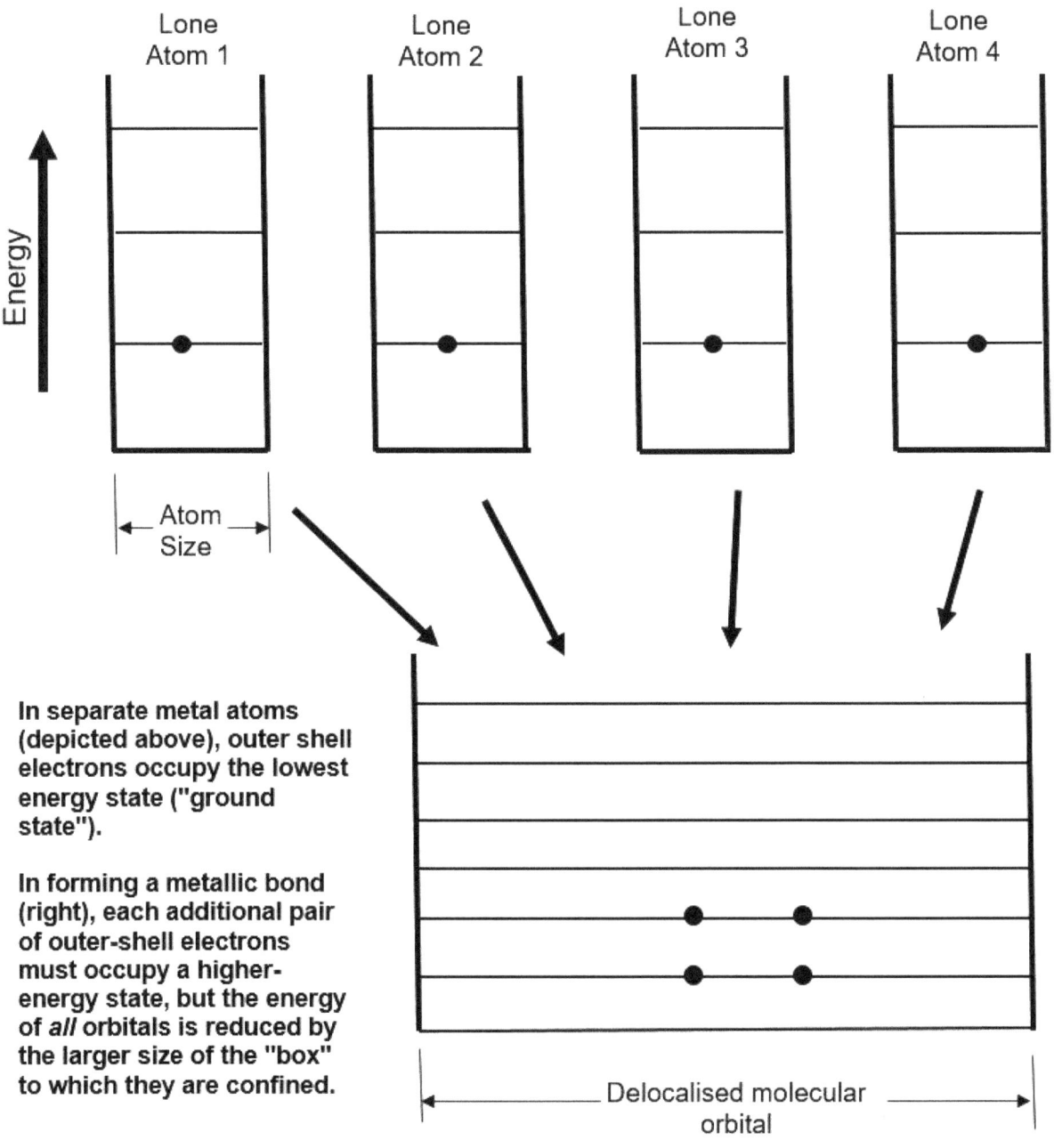

In separate metal atoms (depicted above), outer shell electrons occupy the lowest energy state ("ground state").

In forming a metallic bond (right), each additional pair of outer-shell electrons must occupy a higher-energy state, but the energy of *all* orbitals is reduced by the larger size of the "box" to which they are confined.

9. Chemical reactions

We have seen that the reaction of hydrogen atoms to form a hydrogen molecule occurs because this reaction takes each atom towards having a complete outer shell of electrons, which gives a lower state of energy. However, if other elements or compounds are present that can react, such a reaction may provide a "better deal" for each atom to achieve a complete outer shell, and a state of lower energy.

In particular, as we look across the periodic table (from Column 1 on the left, to Column 7 near the right), the atoms show an increasing tendency to gain electrons. Elements with a nearly-full outer shell of electrons (such as oxygen and fluorine) have a strong affinity for electrons. These elements are non-metals.

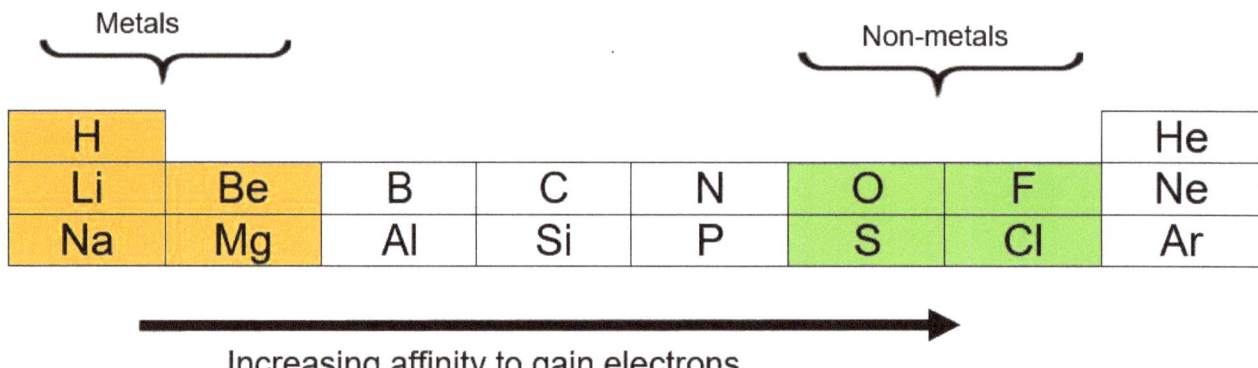

Increasing affinity to gain electrons

By contrast, elements with one or two electrons in their outer shell (like lithium, beryllium, sodium and magnesium) hold their outer-shell electrons very weakly. These elements are metals. Because their outer-shell electrons are held so weakly, these electrons can drift freely from atom-to-atom when the metal is in the solid or liquid state. Since the electrons are not bound to particular atoms, but can drift to neighbouring atoms, metals are good conductors of electricity. These "free" electrons also scatter light that hits the surface of metal, and are responsible for the shiny (specular) surface of metals.

Let's consider what happens when a molecule of hydrogen (H_2) encounters a molecule of oxygen (O_2). The reaction of hydrogen with oxygen would enable the outer shell electrons to be shared between hydrogen and oxygen atoms, but with the electrons concentrated more around the oxygen atom, which has a higher affinity for electrons. This offers a lower state of energy for all the atoms involved. Consequently, energy is released during the reaction of hydrogen with oxygen.

$$2H_2 + O_2 \rightarrow 2H_2O + \text{energy}$$

In order for hydrogen to begin to react with oxygen, a source of energy must supplied to sufficiently weaken or break the existing hydrogen-hydrogen or oxygen-oxygen bonds. The initial "activation energy" can be provided by a small flame or spark. Once the reaction begins, heat liberated by the reaction will raise the temperature of the surrounding gas and provide the activation energy for more hydrogen and oxygen to react.

The reaction of hydrogen and oxygen can be quite explosive, as shown in the following video, in which hydrogen-filled balloons are ignited with a small flame:

>Reaction of hydrogen and oxygen, Ignition of hydrogen balloon, 1 minute
>https://www.youtube.com/watch?v=DbmV5lM4z74   (l = lower case L)

During the explosive reaction of hydrogen and oxygen, the energy of reaction is released as heat. However, the reaction can be undertaken within an electrochemical fuel cell, with the energy released by the reaction being used to drive electrons through an external circuit. In this case, the energy of reaction is converted directly into electrical energy, which can be used – for example – to drive an electric motor and power a car. The concept of hydrogen-powered vehicles has been researched by major car manufacturers, and most have developed prototype vehicles using fuel cells to convert hydrogen fuel into electricity. The city council of Perth has trialled hydrogen-powered fuel cell buses.

The reaction between hydrogen and oxygen to form water can also be undertaken in reverse, with the energy provided by an external source of electrical energy (a battery, for example). The reaction is the opposite of the one given previously:

$$2H_2O + Energy \rightarrow 2H_2 + O_2$$

The decomposition of water into hydrogen and oxygen, using an electrical source of energy, is termed "electrolysis". Electrolysis can be undertaken with very simple equipment, as shown in the following video:

>Electrolysis of water, simple set-up, 3 minutes
>https://www.youtube.com/watch?v=HQ9Fhd7P_HA

>Another short video showing the electrolysis of water is:
>https://www.youtube.com/watch?v=OTEX38bQ-2w

In this latter video, the two gases produced by electrolysis are tested to show that these are hydrogen and oxygen. These are classical, simple tests:

- A test tube of hydrogen produces a characteristic "pop" when ignited by a glowing wooden splint. This is done in a **dry** test tube, so that you can see droplets of liquid water that are formed by the reaction and condense on the walls of the test tube (look carefully during the video).

- A glowing wooden splint thrust into a test tube of oxygen gas will immediately burst into flame. Anything which burns in air (which contains 21% oxygen) will burn far more vigorously in pure oxygen (a fact overlooked in the early design of the Apollo spacecraft, leading to the death of three astronauts in a fire during a training exercise).

Even more vigorous than the reaction of hydrogen with oxygen, is the reaction of hydrogen with chlorine. Chlorine is a strong non-metal, being in Group 7 of the periodic table (with 7 electrons in the outer shell). All Group 7 elements have similar properties, and are termed "halogens". Once again, for this reaction, the existing chemical bond holding the chlorine molecule together must be weakened or broken before the reaction will start. The activation energy for the reaction can be provided by heat, a spark or by ultraviolet radiation.

The following video shows that the reaction of hydrogen and chlorine is not initiated by red, yellow, green or blue light (which doesn't have enough energy to break the chlorine-chlorine bond), but occurs explosively when initiated by ultraviolet light.

> Reaction of hydrogen with chlorine, UV initiated, 3 minutes
> https://www.youtube.com/watch?v=NN82GoBG98s

From an energy viewpoint, the reaction of hydrogen with chlorine can be viewed as shown in the diagram below. For the reaction to start, some energy source (heat, light, etc) must initially provide the "activation energy" $E_{activation}$ to weaken or break the chemical bond in the chlorine molecule. Once free chlorine atoms are formed, these can then react with hydrogen molecules, and – from an energy viewpoint – the reaction will run "downhill" from there. Hydrogen chloride molecules (HCl) formed by the reaction are at a much lower energy state than the initial reactants, with the reaction liberating the Energy of Reaction $E_{Reaction}$.

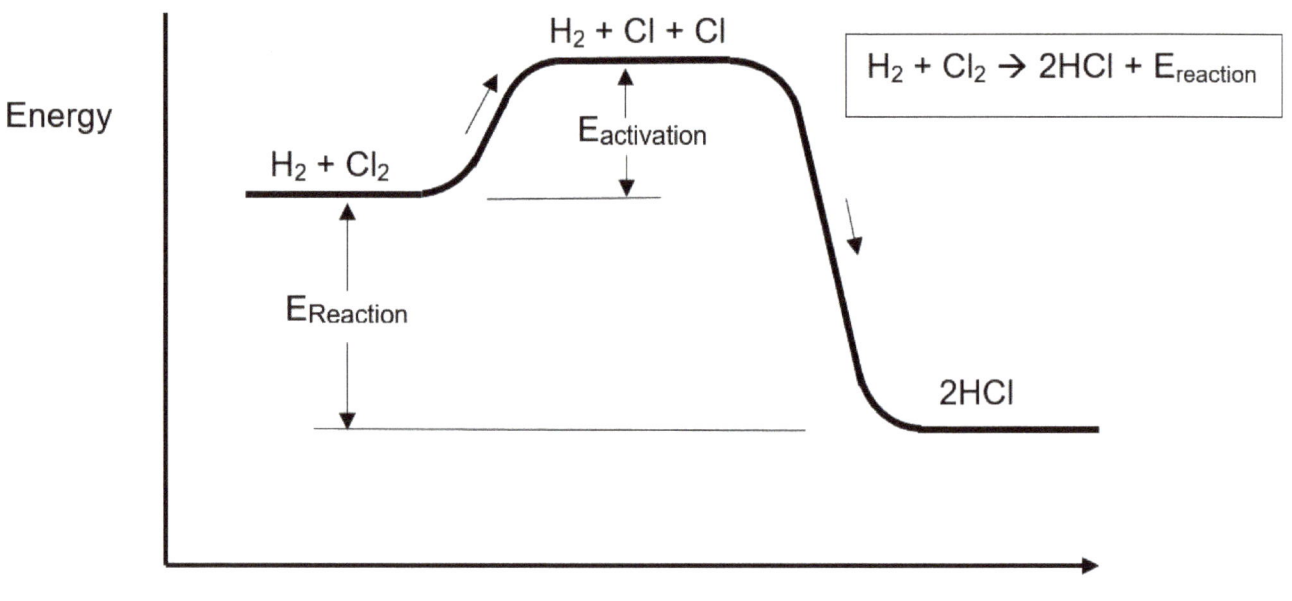

The halogens (Group 7 elements) have the strongest affinity for electrons, and fluorine has the highest electron affinity of any element. Fluorine will react violently with any metal . . . or almost any other element or compound. The reaction of fluorine with sodium metal is shown in the following video:

> Reaction of sodium metal with chlorine, 1 minute
> https://www.youtube.com/watch?feature=player_embedded&v=Mx5JJWI2aaw

Fluorine is so reactive, that it will react spontaneously with many other elements, even with no heat, light or other ignition source.

> Reactions of fluorine with other elements, spontaneous ignition, 1 minute
> https://www.youtube.com/watch?v=mG6EG_igTGw

How can we explain the high electron affinity of fluorine, and more generally, how can we explain the increasing electron affinity for elements as we go from lithium (with one electron in its outer shell) across to fluorine (with 7 electrons in its outer shell)?

Consider the situation of an outer-shell electron - let's call her Linda - in a lithium atom. Let's compare Linda's situation with that of her cousin Heidi residing in a hydrogen atom. The comparison goes like this:

- Heidi's situation is simple. Heidi feels the influence of the one proton in the nucleus of her hydrogen atom, and is attracted by its electrical charge of +1.

- Linda looks towards the nucleus of her lithium atom. The charge in the nucleus of the lithium atom is +3, but Linda doesn't "feel" the full effect of this charge. The positive charge of the nucleus is partly shielded by the two electrons in the inner shell of the atom. The two inner-shell electrons reside in orbitals that are pulled close to the nucleus, with about 85% of their charge inside Linda's outer-shell orbital. So, the "Effective nuclear charge" that Linda feels from the nucleus is as follows:

    Effective nuclear charge = +3 − (2)(0.85) = +1.3

Thus, it "feels" to Linda like the nucleus of her lithium atom has a charge of +1.3 protons.

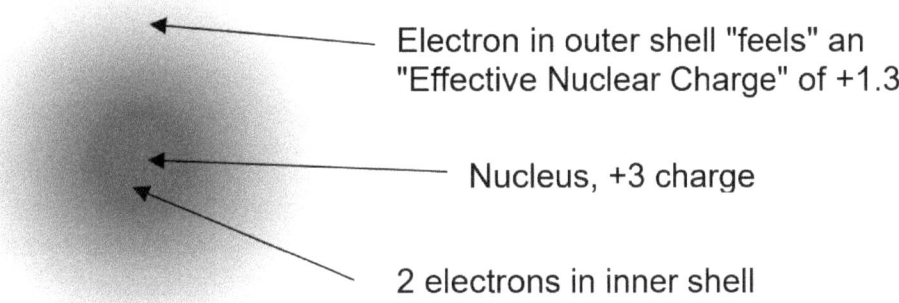

Now comes the interesting bit. How does Linda's situation compare with the situation of her sister-in-law Florence, who resides in the outer shell of a fluorine atom?

Florence looks towards the nucleus of her fluorine atom, where there are 9 protons, with a charge of +9. Fluorine, like lithium, has two inner shell electrons, of which 85% of their charge is inside the outer shell. However, Florence (unlike her sister-in-law in the lithium atom) occupies the outer shell of her fluorine atom with six other electrons. The negative charge of these outer-shell electrons will tend to shield the positive charge of the nucleus, but not nearly as much as electrons in the inner shell. **For electrons within the same shell, only about 35% of their charge is effective in shielding the positive charge of the nucleus**. Since Florence occupies the outer shell of a fluorine atom with six other electrons, the Effective Nuclear Charge that Florence "feels" is:

    Effective nuclear charge = + 9 − (2)(0.85) − (6)(0.35) = +5.2

Thus, it "feels" to Florence like the nucleus of her fluorine atom has a charge of 5.2 protons. The attractive force pulling Florence towards her nucleus is **four times** as great as the force pulling her sister-in-law Linda towards the nucleus of her lithium atom. As a result, Florence (and her fellow outer-shell electrons) will be pulled in and held very close to the nucleus. In fact, the diameter of a fluorine atom (that is, the diameter within which there is a very high probability of finding an outer shell electron) is one-quarter that of a lithium atom. The size of atoms reduces progressively from lithium to beryllium, to boron, etc as we go from Group 1 to Group 8 of the periodic table. For the same reason, the energy required to remove an outer shell electron from the atom (its "ionization potential") increases progressively from Group 1 to Group 8 of the periodic table.

The energy holding an outer-shell electron to a fluorine atom is sixteen times greater than for an outer-shell electron of a lithium atom. Should a fluorine atom get the opportunity to gain an extra electron, it will be pulled in close to the nucleus, in a low state of energy. Consequently, the transfer of an outer-shell electron from lithium (or any metal in Group 1 of the periodic table) to fluorine entails a large release of energy.

So, the affinity for atoms to gain electrons *increases from left-to-right* across the periodic table. But the affinity for electrons *decreases as we go down the periodic table*. So, among all the halogen elements, fluorine is the strongest and most reactive non-metal. However, the ability of metals to lose electrons becomes even greater as we go down the periodic table. The ability of any atom to lose electrons (its "metallic character") increases as we go from hydrogen to lithium, to sodium, to potassium, to caesium.

Because sodium has a stronger tendency to lose electrons than hydrogen, sodium can displace hydrogen where it has reacted with a non-metal. So, if we added sodium to water (hydrogen oxide), the sodium will displace the hydrogen to form sodium oxide. The reaction liberates energy in the form of heat.

$$Na + H_2O \rightarrow Na_2O + H_2 \quad + Energy$$

The sodium oxide (Na_2O) then reacts with water for form sodium hydroxide (NaOH, a strong alkali). In fact, the Group 1 metals are called "alkali metals" because their oxides are alkaline.

When we add sodium to water, we can readily see bubbles of hydrogen being formed, and we can often see steam produced as heat is liberated by the reaction. After a period of time, the heat released by the reaction ignites the hydrogen, often with explosive results. The results become more and more spectacular as the experiment is conducted with lithium, then sodium, then potassium, and then caesium. These spectacular results are shown in the following video, which makes impressive viewing (but don't pay too much attention to the ending, where the film-makers were just being silly buggers, showing stock footage of a nuclear explosion).

> Reaction of alkali metals with water, 3 minute.
> https://www.youtube.com/watch?v=HvVUtpdK7xw

Another impressive example of a displacement reaction, where one metal displaces a "weaker" metal, is the so-called "thermite reaction". As it turns out, aluminium is a surprisingly strong metal, with much greater tendency to lose electrons than does iron. Consequently, aluminium reacts with iron oxide, to form aluminium oxide and iron. This reaction liberates a huge amount of heat – enough heat to melt the iron!!

$$Al + Fe_2O_3 \rightarrow Al_2O_3 + Fe \quad + Energy$$

I have heard that the thermite reaction was used by commandos in the second World War to quietly disable the enemy's trucks. A "thermite bomb" was placed on the bonnet of the truck and, when ignited, the molten iron (at something like 1,500°C) would melt through the metal and the engine of the truck.

Here are some videos of the thermite reaction:

> Thermite reaction, 2-1/2 minutes
> https://www.youtube.com/watch?v=a8XSmSdvEK4

> Thermite reaction, 30 seconds
> https://www.youtube.com/watch?v=CWMATrOatRw

10. The age of metals

The past few thousand years have been characterised as the "iron age" and - although some people refer to the present era as the "information age", "computer age", etc – the "iron age" has not ended. Industrial-scale production of iron and steel underpinned the industrial revolution, and during the nineteenth and twentieth century, a large and efficient steel industry was considered a key requirement for a country to be considered a major industrial economy and a military and world power. Since most conventional weapons and infrastructure (warships, tanks, cannons, bridges, merchant ships, etc) are largely made of steel, a domestic steel industry was essential to the war effort of all major allied and axis powers during the first and second World War.

Metals have unique properties for which they are widely utilised in many applications. Because metals have one, two or three electrons in their outer shell, these electrons are loosely-bound and "delocalised". The outer shell electrons are free to move throughout the metal. This is very different from other elements, in which outer shell electrons are bound to a single atomic nucleus. The presence of "delocalised" electrons is responsible for the unique physical properties which characterise all metals. In particular:

- Metals are ductile and malleable. They tend to bend or deform, rather than shatter, when subjected to impact or force. When a metal is subjected to force, the atomic nuclei can be stretched apart or pushed closer together, and the outer shell electrons can move freely into new positions where they can still bond the atoms together. This malleability has allowed metals be heated and hammered into shape by blacksmiths in earlier eras, and currently, to be hot-rolled into plates or sheets, or extruded through dies.

- Metals conduct electricity and heat. Some metals conduct electricity much better than others, but the electrical conductivity of all metals is generally millions or billions of times greater than glass and ceramic materials.

- Metals are shiny. Their "specular surface" reflects light – usually including visible light, infrared and ultraviolet light. Light radiation, which consists of alternating electric and magnetic fields, causes unbound outer shell electrons to vibrate and re-radiate light at the same frequency. The presence of unbound outer shell electrons also prevents light or radio waves from passing through metals.

The modern industrial age was largely founded on the production of metals on a large industrial scale. Iron and steel alloys were the pre-eminent structural material in the 18th and 19th century. Aluminium, magnesium and titanium become increasingly important in the 20th century for aircraft and other applications requiring high strength and low weight.

However, most metals are hardly ever found as free ("native") metal in the natural environment. For the past three billion years or so, the Earth's atmosphere has been rich in oxygen, ever since algae, bacteria and plants developed the capability to undertake photosynthesis. Due to the high concentration of oxygen in the atmosphere, the surface of the Earth is an oxidising environment. Consequently, most metals in the Earth's crust have been oxidised (that is, have become chemically bound to oxygen or other non-metallic elements). Rocks generally contain metals that are chemically bound to oxygen (but sometimes the metals are bound to sulfur, or are in the form of carbonates or silicates).

To produce metal from the rocks (ores) in which they are present, we need to remove the oxygen (or other non-metals) to which metal atoms are bound. We must reverse the process of oxidation which has occurred, which is called "reduction".

Let's consider how iron and steel are produced. The basic process has not changed in three thousand years, although the scale of production has increased massively, and current technology is much more refined and sophisticated than in the ancient world. Currently, about 1,500 million tonnes of steel is produced each year. This is equivalent to about 230 kilograms being produced each year for every man, woman and child on Earth. Steel production is several times greater than the amount of food produced by humanity.

Iron oxide is widely abundant in the Earth's crust, generally existing as hematite, Fe_2O_3. In this state, the iron exists as Fe^{+3} ions, that is, iron atoms which have each lost three outer-shell electrons to oxygen atoms. To produce iron metal, the iron must be "reduced". Iron oxide must be reacted with something which has an even greater ability to lose electrons and to bind with oxygen. As we have seen previously, we could use a more active metal, like aluminium, to displace the iron (as occurs in the thermite reaction). However, aluminium has a greater economic value than iron, so this would not be practical or sensible. We need a "reducing agent" that is cheap and widely available. Carbon (from coal) is used.

The production of iron

Since the beginning of the industrial revolution, iron has been produced in huge reactors called "blast furnaces". A modern blast furnace would produce about 13,000 tonnes of iron each day. The blast furnace is a vertical cylindrical vessel that is filled with iron ore (sintered into nodules about the size of a golf ball), carbon (coke, also in golf ball-sized nodules) and limestone. Air is blown (blasted) into the bottom of the furnace, so that burning coal provides heat and high temperatures for the process. The bottom of the furnace is heated well above 1,500°C, the melting point of iron. The reaction for the combustion of coke (carbon) is:

$$C + O_2 \rightarrow CO_2$$
Carbon   Oxygen   Carbon dioxide

Once the extremely hot combustion gases rise above the combustion zone at the bottom of the furnace, they come into contact with additional carbon, which "reduces" carbon dioxide to carbon monoxide:

$$CO_2 + C \rightarrow 2CO$$
Carbon dioxide   Carbon   Carbon monoxide

Note that carbon monoxide is very unstable. The carbon atom does not have a complete outer shell of electrons. This reaction absorbs energy, and it only occurs because of the very high temperatures and the ample amount of carbon available to react.

Carbon monoxide is chemically reactive, especially at the high temperatures inside the blast furnace. The carbon atom is desperate to complete its outer shell of electrons, and it can do this by seizing another oxygen atom from iron oxide. Each carbon monoxide molecule gains one oxygen atom, so the reaction is:

$$Fe_2O_3 + 3CO \rightarrow 2Fe + 3CO_2$$
Iron oxide   Carbon monoxide   Iron   Carbon dioxide

Hot liquid iron formed in the reaction is very dense. It dribbles downwards through the furnace and collects at the bottom.

If we add all the above reactions together, the overall reaction for the production of iron is:

$$Fe_2O_3 + 3\,C + 3/2\,O_2 \rightarrow 2Fe + 3CO_2$$

This equation tells us that, if the steel making process were 100% efficient, 3 moles of carbon would be needed to produce 2 moles of iron. Taking into account the relative weights of carbon and iron atoms (their atomic weights), we can easily calculate that at least 325 grams of carbon are burned for each kilogram of iron produced. The iron production process is much less than 100% efficient, and roughly one kilogram of metallurgical coal (coking coal) is used to make each kilogram of iron.

Limestone (calcium carbonate) is added as a "flux". It reacts with sand (silica, SiO_2) and other impurities present in the iron ore or coke to form a liquid "slag" of calcium silicate. The molten slag also trickles down through the furnace and collects at the bottom. Because it is less dense than iron, the slag floats on top of the iron. Periodically, liquid iron and liquid slag are drawn off through ports in the bottom of the blast furnace.

As soon as liquid iron is formed, and as it trickles down through the blast furnace, it comes into contact with carbon. At the high temperatures in the blast furnace, carbon dissolves in liquid iron. The "pig iron" produced in a blast furnace is a saturated solution of carbon dissolved in molten iron. It contains about 4% carbon. As the iron cools, some of the carbon (about 2%) remains dissolved and is contained within the crystal structure of the solid iron, increasing its hardness. The remaining carbon crystallises as graphite particles within the iron, and this makes the carbon brittle and subject to fracture.

Thus, where greater strength is required, iron is converted to steel by reducing its carbon content below about 2%. Various processes have been developed to convert iron into steel. In each process, a stream of air or oxygen is blown into hot liquid iron, oxidising some of the carbon to carbon dioxide. During the steelmaking process, other metals may be added to enhance specific qualities that are desired (tensile strength, corrosion resistance, ease of welding, ease of machining, etc). There are various grades of steel, each produced to particular specifications, with the choice depending upon the requirements and budget of the customer.

A good overview of the blast furnace and steel making process is provided by the ten-minute video which may be viewed at:
   https://www.youtube.com/watch?v=Ea_7Rnd8BTM

A series of six short videos (2-3 minutes each) describing iron and steel production, stainless steel, recycling and applications can be viewed at:
   http://science.howstuffworks.com/blast-furnace-videos-playlist.htm

If you overlook the annoying gung-ho narration, and use of Fahrenheit temperatures, these videos are quite good.

The fate of metals

Once iron and steel (and other metals) are made into bridges, buildings, cars, etc, they are exposed to the air, and will eventually oxidise ("rust"). Basically, the metal returns to the oxidised state in its original ore rock. For metals, this is the equivalent to "ashes to ashes, and dust-to-dust".

Many methods have been developed to enhance the corrosion resistance of metals, and thus, to extend their useful life. Since rusting depends upon contact with air (oxygen), and is greatly accelerated by the presence of water and salts, one of the simplest means of corrosion protection is by covering the surface with paint or other impervious coating. These coatings act as a barrier to prevent oxygen, water and salts from reaching the metal surface.

Different metals and metal alloys vary widely in their susceptibility to corrosion, and – surprisingly - the susceptibility of a metal to rusting is not directly related to how strongly it bonds to oxygen. Unprotected iron and steel are generally very susceptible to rusting. When the surface begins to oxidise, it forms a layer of iron oxide (Fe_2O_3), which is brittle and porous. The iron oxide surface layer easily flakes off, exposing underlying metal to further oxidation and corrosion. Even if the iron oxide layer doesn't flake off, it is porous and allows the passage of oxygen, water and salts to the underlying surface.

The corrosion resistance of steel is greatly enhanced by the addition of chromium and other metals. Such alloys are referred to as "stainless steel". The use of stainless steel pots and pans is now largely taken for granted, but I remember that when I was growing up, stainless steel was considered a wonder material of the modern age. In 1964, I visited the World's Fair in my home town of New York City, which featured a 12-storey-high stainless globe of the world (and it's still there!! See photo here from: http://en.wikipedia.org/wiki/Unisphere). When stainless steel alloys are initially exposed to air, a transparent, strong and non-porous oxide coating forms on the surface. This coating provides an effective barrier that protects the underlying surface from further oxidation.

Source: Flapane at Italian Wikipedia

The same protective mechanism is responsible for the high corrosion resistance of aluminium and magnesium – even though aluminium and magnesium are much more active metals (more readily oxidised) than iron. Like stainless steel, aluminium and magnesium form a strong, non-porous, transparent oxide coating which protects the underlying surface from further oxidation.

Eventually, all products will reach the end of their useful life. Sometimes, they will rust, wear out or become too costly to fix and maintain. In other cases, they will become obsolete and overtaken by improved technology. Or, they might simply fall out of fashion, or fall victim to changing priorities or circumstances. Today's shiny new sports car, with the latest u-beaut wi-fi music and communication system, will ultimately become junk. So, what will happen to it – and to the other 1,500 million tonnes of steel produced this year - in 10, 20 or 50 years from now? The steel and other materials will either be separated and recycled, or they will end up being buried in rubbish dumps.

The good news is that much steel, aluminium and other metals is recycled. Of course, recycling avoids the need to mine new metal ore. Perhaps even more important, recycling requires only a fraction of the energy required to produce metal from ore. The energy required to recycle steel (usually, by melting it in an electric arc furnace) is about **half** that required to produce steel from iron ore. In the case of aluminium, the difference is more dramatic. Recycling aluminium requires **twenty times less** energy than is needed to produce aluminium from ore.

Producing steel and other metals, cement and plastics accounts for a very significant share of mankind's consumption of fossil fuels and greenhouse gas emissions. Steel production requires huge amounts of energy (coal for blast furnaces, electricity for arc furnaces), partly because huge amounts of steel are produced. But aluminium production is far more energy-intensive. To produce each kilogram of aluminium requires about ten times as much energy as to produce the same weight of steel (although aluminium provides more strength per unit weight).

As we progress into the 21st century, lightweight, high-strength metals like aluminium, magnesium and titanium are becoming increasingly important for air transport and other applications. Australia is a major producer and exporter of aluminium, and aluminium production is a major reason why Australia is reputed to have the highest per capita greenhouse gas emissions of any nation on Earth. We'll look at aluminium production in the next chapter.

11. Aluminium and electrolytic refining of metals

As we have seen, most metals occur naturally in the oxidised state, usually combined with oxygen. The metal can be freed by reacting it with a "reducing agent" which has a greater affinity for oxygen, and a greater ability to lose electrons. In many cases, carbon (or coke, made from coking coal) is used as a reducing agent. However, some metals have a very strong tendency to be oxidised (a high "oxidation potential"). These metals have a stronger affinity for oxygen than carbon, and cannot be reduced with carbon or any other common reducing agents. One important metal of this type is aluminium.

About 40 million tonnes per year of aluminium are produced. This is equivalent to about 6 kilograms for each person on Earth. Australia is a major producer and exporter of aluminium.

To produce aluminium metal, we need to reverse the oxidation process that has occurred naturally.

$$2Al + 3/2 O_2 \rightarrow Al_2O_3 + \text{Energy}$$
$$\text{Aluminium} \quad \text{Oxygen} \quad \text{Aluminium oxide}$$

Oxidation of aluminium releases a considerable amount of energy, about 65 megajoules per kilogram of aluminium. The oxidation reaction can be driven in reverse if we provide an energy input of at least this amount, and this is exactly what is done when aluminium is produced in an electrolysis cell.

An external source of electricity (usually, a coal-fired power station or hydroelectric plant) provides the energy required to decompose aluminium oxide (alumina) into aluminium metal and oxygen. The process occurs in an electrolytic cell, in which aluminium oxide is dissolved in hot molten cryolite (sodium aluminium fluoride). The inside of the cell is lined with an electrode (cathode) made of carbon, and another carbon electrode (anode) is lowered into the molten cryolite bath. An electrical current is passed between the two electrodes. The cell is maintained a temperature of about 1,000°C, which is above the melting point of the cryolite and above the melting point of aluminium metal.

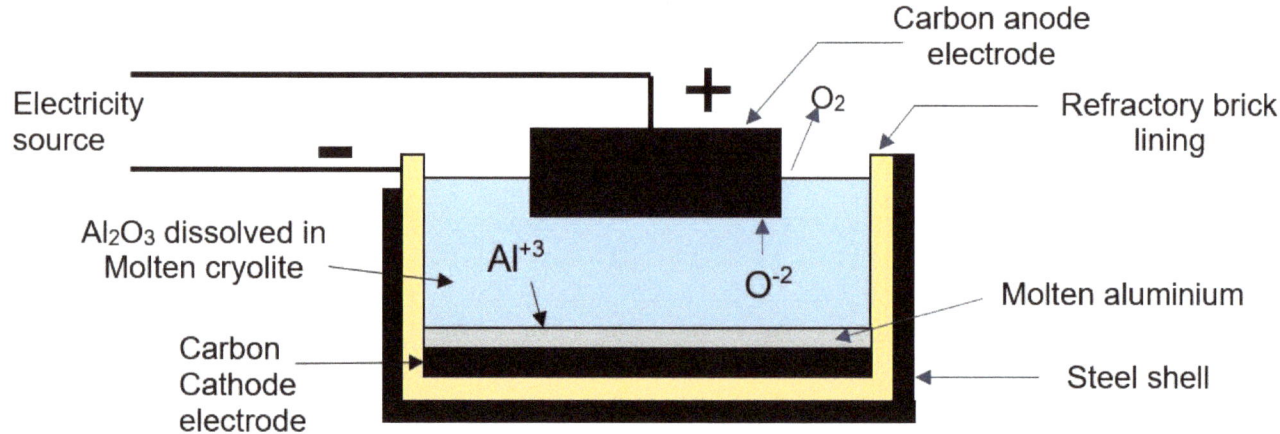

Aluminium oxide dissolves in molten cryolite, producing free aluminium ions (Al^{+3}) and oxide ions (O^{-2}). The electrical voltage applied to the cell causes aluminium ions to be attracted to the negative electrode, where each Al^{+3} ion gains 3 electrons and forms an atom of aluminium metal. Molten aluminium, being denser than the cryolite electrolyte, collects at the bottom of the cell and is periodically drawn off.

Meanwhile, oxide ions are attracted to the positive electrode, where each O^{-2} ion loses 2 electrons and forms a neutral oxygen atom. Most of these oxygen atoms form oxygen molecules (O_2), and oxygen gas bubbles off the positive carbon electrode. Some oxygen atoms react with the carbon electrode to form carbon dioxide gas.

The reactions occurring at the electrodes are as follows:

$$\text{Cathode:} \quad Al^{+3} + 3e \rightarrow Al$$
$$\text{Anode:} \quad O^{-2} - 2e \rightarrow O_2$$

Notice that a **gain of electrons** occurs at the cathode, where Al^{+3} ions are *"reduced"*. A **loss of electrons** occurs at the anode, where oxide ions are *"oxidised"*.

Note that "oxidation" *always* involves the loss of electrons, and "reduction" *always* involves the gain of electrons. The two processes, oxidation and reduction, always occur together in chemical reactions. Whenever something is oxidised, something else is reduced.

A very useful pneumonic to remember is:
"**LEO** the lion says **GER**"

For **LEO**: **L**oss of **E**lectrons is **O**xidation
For **GER**: **G**ain of **E**lectrons is **R**eduction

A good 5-minute video, which shows the production of aluminium and explains the process, can be viewed at:
    https://www.youtube.com/watch?v=NW1k4wNEq14   (1 = number 1)

The production of aluminium is very energy intensive. To provide the energy required for the reaction, at least 65 megajoules of electrical energy must be provided for each kilogram of aluminium produced. When electricity is produced from coal or other fossil fuel, about 3 units of fuel energy is consumed for each unit of electrical energy produced. Furthermore, various losses occur in the electrolytic cell. According to one recent article, an average of 211 megajoules of energy is used to produce each kilogram of aluminium. This corresponds to burning 8 kilograms of coal for each kilogram of aluminium produced.

Electrolytic production of other metals

Very similar electrolytic processes are used to produce other highly active metals, such as magnesium (also used in lightweight alloys for aircraft and cars) and sodium (used in specialty applications). In each of these cases, an external source of electrical energy is used to drive the chemical reaction that converts the metal from its oxidised state (such as Al^{+3}, Mg^{+2}, Na^+) to its reduced state.

Another 5-minute video shows the electrolysis of molten salt (sodium chloride) to produce sodium metal:

https://www.youtube.com/watch?v=NinmIYKaj2w   (I – Capital letter i)

Electroplating of metals

Similar electrolytic processes are used to electroplate metals. For example, metal jewellery may be coated with gold for decorative effect and to avoid tarnishing or rusting. In other cases, metal parts are coated with other metals for corrosion resistance, wear resistance or decorative effects.

Electroplating can be done very simply, using very simple equipment, as is demonstrated in a one-minute video:
    https://www.youtube.com/watch?v=FnJ0V7B7nKo

To produce good quality electroplated coatings is more complex and requires specialised knowledge. The surface being coated must be immaculately clean, and various chemicals are often added to the electrolyte solution. Sometimes the temperature or other conditions may need to be strictly controlled. A five-minute video showing commercial electroplating of metal parts can be viewed at:
    http://www.aft-corp.com/processes

Electrolysis of copper

In the case of copper, very high purity is required to provide high electrical conductance for the manufacture of wire and cable. Usually, copper ores are chemically reduced in a furnace, producing "blister copper", at about 99% purity. This impure copper is relatively brittle, and not suitable for many applications. In particular, because of impurities, blister copper has poor electrical conductance and is not suitable for use as an electrical conductor.

The impure copper is formed into an anode electrode which is immersed in a solution of copper salts, while a cathode electrode of pure copper is immersed into the same solution. Electrical current is passed through the electrochemical cell. The electrical current causes copper metal at the surface of anode to be oxidised, forming Cu^{+2} ions, which go into the solution. The copper anode literally dissolves during the process. At the same time, Cu^{+2} ions in solution are reduced to copper metal and electroplated onto the cathode. The cathode literally grows in size as electrical current passes through the cell.

As the copper anode oxidises and dissolves in the solution, metal impurities are either released to form a sludge at the bottom of the cell (for metals like silver, with a lower "oxidation potential" than copper) or remain in solution (for metals like zinc, which are more readily oxidised than copper).

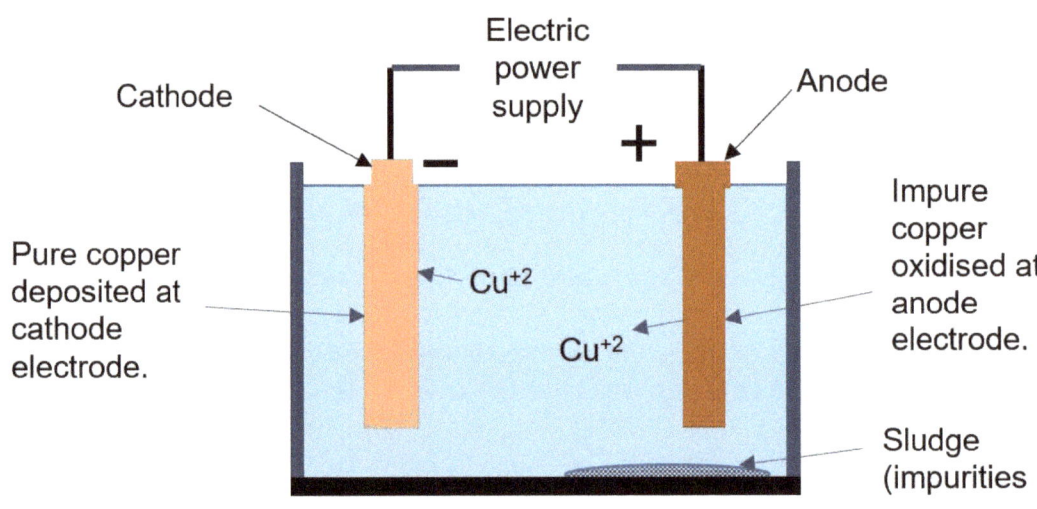

Eventually, the electrodes are removed from the cell:

- The pure copper cathode is melted: some is used to form a new cathode for the next electrolysis step, and the remainder is taken away to be extruded into wire, cable or tubing.
- The shrunken anode is mixed with more impure "blister copper", melted and formed into a new anode.

The reactions occurring in the cell are:

Cathode: $Cu^{+2} + 2e \rightarrow Cu$ (Copper ions are reduced)

Anode: $Cu - 2e \rightarrow Cu^{+2}$ (Copper metal is oxidised)

Overall, there is no net reaction for the cell! Copper is simply transferred from one electrode to the other - but the impurities are not! In principle, virtually no energy is required to drive the reaction (that is, the "back-voltage" of the cell is very low), but in practice, significant energy losses occur when large electrical currents are passed through the solution.

12. The transport revolution

A short history of the world, in my experience

Perhaps the most profound social changes that we have seen in the last 75 years resulted from the ubiquitous role of the automobile in shaping our cities, our societies, our economy and culture.

Up to the period before (and during) the Second World War, the typical family did not own – and could not afford – a car. People relied on public transport, bicycles and walking to get to work, school, services and entertainment. Many of the services that are currently centralised in shopping malls were provided by corner shops, local cinemas and deliveries of groceries. Sports were often played by groups of kids from the local neighbourhood in a local park, rather than in formal matches with teams or schools on the other side of the city. Special facilities - like entertainment and sporting complexes – didn't exist or were visited only on special occasions.

This all started to change after the Second World War. In the US, where I grew up, huge industries had grown for mass production of tanks, aircraft, ships and other military equipment. Manufacturing companies and factories needed to shift production from the military to the household consumer market. In an unprecedented period of post-war prosperity, industry produced refrigerators, washing machines and toasters on a scale not seen before. But the pre-eminent product of this era was the automobile. For the first time, cars were produced on a massive scale and at a price that made them accessible to ordinary families.

The wide availability of cars allowed families to move to the outer fringes of cities, into new suburban developments which sprawled outwards. This was facilitated by construction of a vast networks of roads across the United States, including the Interstate Highway system. Just beyond the eastern fringe of my home town of New York City, new residential developments, highways, shopping centres and carparks spread across vast tracts of Long Island (which previously were farms and country estates for the wealthy – as depicted in the movie 'The Great Gatsby'). The same type of suburban sprawl also spread across the Hudson River into New Jersey, with development extending along an almost unbroken corridor that merged into Philadelphia 160 kilometres to the south. Although some public transport was later established to service these areas, the low population density of these sprawling suburbs meant that, for most people, having a car was essential for getting to work, school, shopping and other day-to-day activities.

At the same time, social pressure for people to buy automobiles was provided by extensive advertising, which the new medium of television was bringing into every home. Driving a car was promoted as the ultimate status symbol and expression of individual freedom and individuality. I recall being bombarded by endless advertisements on television, another technology finding its way into every house. These ads invariably depicted a young man driving the latest model car along peaceful and scenic country roads, with a beautiful woman sitting alongside, her blond hair blowing freely in the wind of an open convertible. The musical jingo "See the USA in your Chevrolet" became etched in my brain over hundreds (maybe thousands) of repetitions during the years of my childhood. The message was clear to the young men who had just returned from the hardships of war: "you too could have unlimited freedom to go anywhere you like, be the envy of your mates, and get that gorgeous girl you've been dreaming about". The car was not depicted as a useful transport device, but as a "dream machine" – the key to a life of freedom, romance and affluence.

However, by the 1960s, it was already becoming very clear that the reality of driving in cities and sprawling suburbs looked very different from idyllic scenes shown in ubiquitous car advertisements. In particular, several huge issues were emerging as more and more cars were being driven by more people, more regularly and over longer distances:

1. Emissions of air pollution degraded air-quality and quality-of-life in major cities. Gasoline-powered vehicles produce carbon monoxide, nitrogen oxides and unburned hydrocarbons, while diesel vehicles produce nitrogen oxides and fine soot particles. It was discovered that sunlight caused nitrogen oxides to react with hydrocarbons in the atmosphere to produce a semi-permanent haze of "photochemical smog". Some cities (including Los Angeles) are located in geographical sites which are subject to "temperature inversions", which trap pollution above the city during some climate conditions. This poses a major health hazard for people with asthma and respiratory conditions.

   Over the past 45 years, governments introduced increasingly stringent regulations limiting the amount of pollutant gases which could be emitted, and air quality of many cities in Europe, the United States and Australia has improved (or at least, not become worse). However, with the rapidly growing number of cars and heavy vehicles in China and other developing countries, air quality in these cities has deteriorated markedly (and now rank as the worst in the world).

2. The United States, Europe and other countries became highly dependent on oil imports to meet the fuel requirements of their ever-increasing numbers of cars, trucks and buses. Most major oil exporting countries were (and still are) located in the Middle East and other politically unstable areas. These countries formed a cartel called OPEC (Oil Producing and Exporting Countries). In 1973, and later in 1979, OPEC countries restricted oil supply in response to Arab-Israeli wars, causing severe oil supply shortages and sharp increases in the price of crude oil. These "Oil Crises" highlighted the vulnerability of countries that relied heavily on oil imports to meet their requirements for transport fuels.

3. The congestion of roads and highways led to growing pressure on governments to build additional roads and highways. Enormous costs were incurred to construct roads, freeways, overpasses, tunnels, bridges and carparks, and huge areas of formerly productive land have been covered in asphalt and concrete. However, decades of experience have shown that construction of additional roads encourages more people to own cars, and to drive further. The volume of traffic carried by new, larger roads tends to gradually increase until congestion returns to previous levels.

4. Widespread adoption of cars exacerbated the huge toll of deaths and injuries from traffic accidents. At the peak of the Vietnam War in the late 1960s, more Americans were dying in road accidents (1,000 per week) than were killed in the war. However, in the 1970s, regulations were enacted by governments in the US, Australia and other countries to introduce basic safety features (padded dashboards, collapsible steering columns, no protruding objects to impale pedestrians). These requirements were later extended to include mandatory seatbelts, front and rear "crumple zones" to absorb impact energy during a crash, airbags and other features.

These changes in vehicle design and fit-out, as well as enforcement of seat belt wearing and drink driving penalties in Australia, were remarkably effective in reducing deaths and injuries from car crashes. The number of Australians killed in road accidents is now only one-third that occurring 30 years ago, despite three times as many vehicles operating on the roads. Even so, 23 people die each week in road accidents in Australia (and 600 die each week in the USA), and motor vehicle accidents remain a leading cause of death and injury among young, healthy people.

5. The shape of cities, towns and communities was changed dramatically by the widespread availability of cars. Residential housing sprawled outwards, enabling a new style of suburban living. The low population density of these sprawling suburbs largely prevented the development of effective public transport. Shops, cinemas and other services became larger and centralised within enormous shopping malls.

Can new technologies rescue the future?

By the 1960s, the huge adverse impacts of motor cars was becoming evident, and several new technologies came under consideration to ultimately replace the gasoline and diesel-powered internal combustion engine as a power source for cars and other vehicles. In particular, fuel cells vehicles, electric vehicles and hybrid vehicles promised to provide very effective solutions to dramatically reduce or eliminate urban air pollution and avoid dependency on oil imports (problems 1 and 2 above).

Development of these new technologies has now been underway for about 50 years. Researchers have been attempting to overcome technical, cost and logistical issues that have prevented fuel cell and electric vehicles from taking over the role played by gasoline and diesel-powered vehicles. One challenge for any such technology development program is that it must hit a moving target. Fuel cell or electric vehicles must provide a practical, cost-effective alternative to gasoline and diesel-powered vehicles. These conventional technologies have continually been refined and improved over the years, largely in response to government-mandated requirements on pollution emissions, fuel economy and safety. The level of performance that fuel cell and electric vehicles must achieve to compete against conventional engine technology has been rising. Nonetheless, there is some reason to believe that fuel cell, electric and hybrid vehicles will play a major role in the next phase of the transport revolution.

One factor driving change in the transport sector is the depletion of known sources of conventional crude oil. While conventional crude oil is gradually being replaced with shale oil, tar sands, deep off-shore deposits and other "non-conventional" sources of crude oil, these new sources are more expensive to produce and more environmentally damaging. Technology that allows other energy sources to be used in transport applications would be desirable to circumvent fuel price increases, reduce oil imports and the resulting balance-of-payments shortfalls, and minimise Australia's vulnerability to fuel supply disruptions. Fuel cells and electric vehicles could provide exactly such a diversification of energy supply sources.

The big advantage for fuel cell and electric vehicles

Fuel cell vehicles could provide an ideal solution to reduce or eliminate air pollution in congested cities. Fuel cells produce electrical power by reacting a fuel with air in an electrochemical cell. Because the energy of the reaction is converted directly into electricity, fuel cells are not "heat engines" and are not subject to efficiency limitations that apply to heat engines. Most fuel cells use hydrogen as a fuel, which is reacted with oxygen from the air. The only product of the reaction is water vapour. Fuel cells do not produce the toxic and environmentally harmful pollutants produced by conventional engines.

Hydrogen for fuel cells does not occur naturally in the Earth's crust, but can be produced from virtually any energy source – including coal, natural gas and biomass (woody and agricultural wastes). Coal and natural gas are hydrocarbons (comprised of carbon and hydrogen atoms), while biomass consists mainly of cellulose (comprised of carbon, hydrogen and oxygen). There are three alternative options to produce hydrogen from these energy sources:

1. One method is by gasifying coal, natural gas or biomass, whereby these hydrocarbons undergo incomplete combustion with a small amount of oxygen.

    For example, natural gas consists mainly of the hydrocarbon methane, and the chemical reaction for gasification of natural gas can be shown as:

    <u>Gasification</u>
    $$CH_4 + \tfrac{1}{2} O_2 \rightarrow CO + 2H_2$$
    Methane Oxygen Carbon monoxide hydrogen

    The gasification reaction produces a mixture of carbon monoxide and hydrogen gas, called "synthesis gas". To maximise the production of hydrogen, the "synthesis gas" can subsequently be reacted with steam (water vapour) at high temperature in a "shift reaction". Carbon monoxide is a strong reducing agent, and takes oxygen away from water molecules, to produce additional hydrogen.

    <u>Shift reaction</u>
    $$CO + H_2O \rightarrow CO_2 + H_2$$
    Carbon Monoxide Steam Carbon dioxide Hydrogen

    The overall reaction for gasification of natural gas, followed by a shift reaction, is:

    $$CH_4 + \tfrac{1}{2} O_2 + H_2O \rightarrow CO_2 + 3 H_2$$

    Both the gasification process and shift reaction entail significant energy losses, so perhaps only about half the combustion energy contained in the original hydrocarbon fuel is retained in the hydrogen product. Also, gasification and the shift reaction are only practical in large industrial-scale plants, so the hydrogen would need to be transported to dispersed hydrogen refuelling stations. Pipeline infrastructure to transport and distribute hydrogen to many end-users does not currently exist, and would be expensive to construct.

2. A second route to produce hydrogen is by burning coal, natural gas or biomass in an electrical generating plant, and then using the electrical energy to electrolyse water into hydrogen and oxygen.

    Fossil fuel power plants already exist, as does electricity distribution systems, so most of the infrastructure is already in place to produce hydrogen in this way. Electrolysis of water can be undertaken on a small scale, on-site at refuelling stations. However, generation of electricity in a coal or gas-fired power station is only about 35% efficient, and electrolysis is generally about 80% efficient, so the hydrogen product contains less than 30% of the combustion energy in the original fuel.

3. A third option is to generate electricity using solar, wind or other renewable energy source, and then using the electrical energy to electrolyse water. Solar energy production could be undertaken either at a central solar power station, or at the refuelling station. This would certainly appear to be the best, most efficient and sustainable long-term option for producing hydrogen for fuel cell vehicles.

"Well-to-wheel" energy conversion efficiency

All of these routes rely on a chain of steps, and each step entails a loss of energy and efficiency. Consider, for example, that coal is burned in a power station to generate electricity, and that the electricity is used to electrolyse water (2nd method above). The hydrogen is then stored on-board the vehicle, and is converted to electricity in a fuel cell. Finally, the electricity is converted to mechanical power in an electric motor to drive the wheels of a car.

Generation of electricity from coal is about 33% efficient (converting one-third of the fuel energy in coal to electrical energy), electrolysis is typically about 80% efficient, and a fuel cell would typically be 50% efficient. An electric motor converts about 90% of the electrical energy input to mechanical power. So, the overall "well-to-wheel" efficiency is about 0.33 X 0.80 X 0.50 X 0.90 = 12%. So, only about 12% (roughly one-eighth) of the energy in the coal is ultimately converted to mechanical power driving the wheels of a fuel cell vehicle.

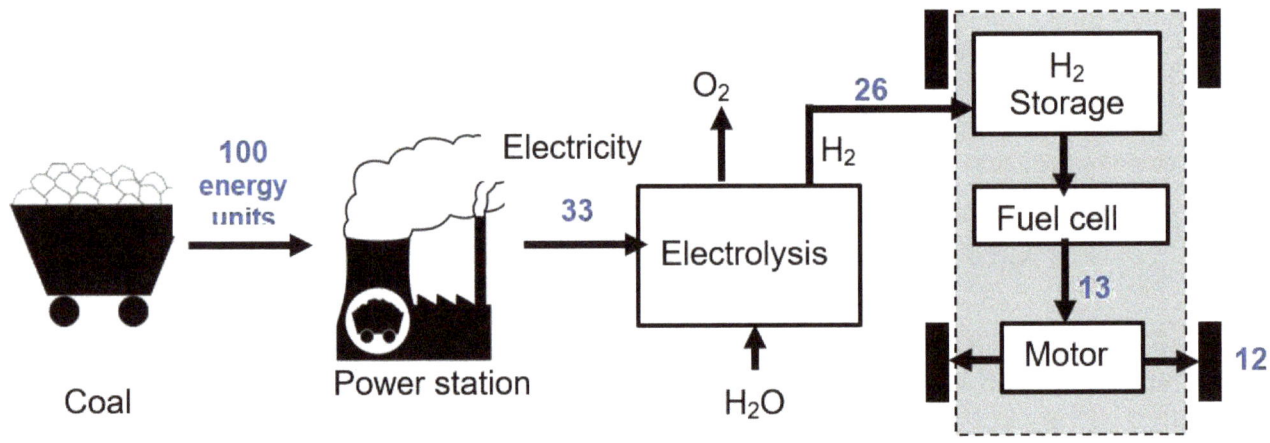

This sounds like a grossly inefficient use of the energy in coal to drive a car, but it is actually not much different from the "well-to-wheel" efficiency of driving vehicles with gasoline or diesel-powered engines. About 10-20% of the energy in crude oil is consumed in its production, transport and refining. Let's consider that the production of petroleum fuels is 85% efficient. The efficiency of a gasoline engine is about 33% at its optimal rated load, but is only perhaps half as much (say, 12%) under typical driving conditions. Power is transferred to the wheels through a transmission, driveshaft and differential, which transmits perhaps 90% of the power. So, the overall "well-to-wheel" efficiency is about 0.85 X 0.12 X 0.90 = 9%. Furthermore, the engine, transmission, driveshaft and differential of a gasoline-powered car account for a sizeable share of its weight (and available space). Potentially, a fuel cell and electric motor would be smaller and lighter than the drive-chain of a conventional car, so less energy might be required to move the car around.

Electric vehicles could provide an alternative option to use coal, natural gas, biomass or renewable energy as an energy source for vehicles. Here too, an energy source like coal could be burned in a power station to produce electricity at about 33% efficiency. Electrical energy would be stored in batteries on-board the vehicle, with about 80% of energy used to charge the battery being released during discharge. Then, once again, an electric motor would convert electricity to mechanical power to drive the wheels at about 90% efficiency. The overall "well-to-wheel" efficiency would thus be 0.33 X 0.8 X 0.9 = 23%. Thus, electric vehicles could potentially have about twice the overall "well-to-wheel" fuel conversion efficiency of conventional gasoline-powered cars. But to achieve this high energy conversion efficiency, the battery technology must be able to provide the required energy storage density to give the range and performance that is needed.

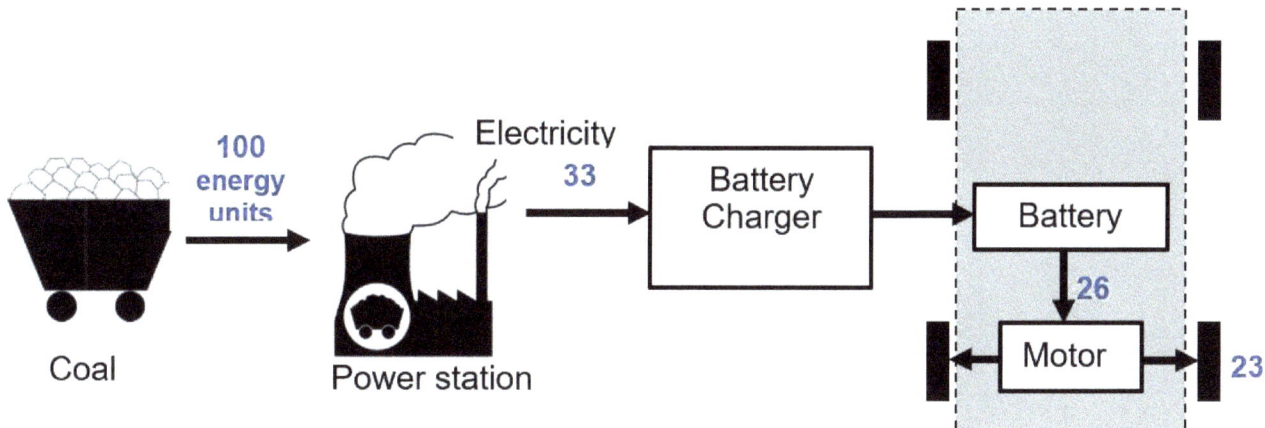

Fuel cell and electric vehicles can also play a role in reducing air pollution in major cities. If the energy source for electricity or hydrogen production is solar or wind energy, then clearly, fuel cell and electric vehicles could be virtually pollution-free. But even if the energy source is coal, then this could still improve air quality in our cities because the electricity generation or gasification plant would generally be located well outside the airshed of major cities. In effect, fuel cell and electric vehicles provide a means to shift air pollution from cities to areas with low populations.

13. Electrochemical Batteries

We have previously seen that we can use electrical energy to drive a chemical reaction that would otherwise not occur. We can also do the opposite: we can use energy released by a chemical reaction to generate electricity. This is done in electrochemical "batteries" that operate our mobile phones, laptop computers and tablets, toys and many other portable devices, and may in future be used to propel our cars.

An assortment of single-cell "batteries", including 1.5-volt alkaline primary and rechargeable cells and a 3.7 volt lithium ion camera battery.

A "battery" contains one or more electrochemical cells, each cell using a chemical reaction to impart energy to electrons flowing through our device. Each cell typically generates 1-2 volts, with cells often connected in series to produce voltages of 6, 9, 12 or higher voltages.

The term "battery" was originally used to describe a collection of artillery cannons used by armies to send a barrage of cannon balls at the enemy. At the lower end of Manhattan, in my home town of New York, is Battery Park, named after a fort which had a battery of 26 cannons intended to protect New York harbour from attack by the British fleet during the War of 1812 (as it turned out, the British did not attack New York, but they did bombard Washington DC and then Philadelphia). The fort at New York's Battery Park is still there, but the cannons are long gone.

Up until about 1950, electrochemical batteries often consisted of many cells (a "battery" of cells) connected in series to provide high voltages. At that time, radios and other electronic devices used vacuum tubes, which needed high voltage to operate – up to several hundred volts. Now, electronic devices with transistors and integrated circuits need only a few volts to operate. Most household devices use AA or AAA-size batteries, each battery containing a single cell producing about 1.5 volts. It is a quirk of history that we still call these "batteries", even though they contain only one electrochemical cell.

The chemical reaction occurring inside a battery produces a voltage (originally called an "electromotive force") to "push" or drive electrons through a circuit. A good analogy is the flow of water from a reservoir at the top of a hill. When we open a tap at the bottom of the hill, water shoots out under pressure (corresponding to the voltage provided by a battery). We could use the stream of pressurised water to blast the mud off our car, drive a turbine, or do useful work in other ways. The pressure of the water is derived from the force of gravity acting over the height of the reservoir above the tap (that is, the "gravitational potential energy" of water in the reservoir). The pressure is a measure of the energy expended per unit volume of the water.

In a similar way, electrons in the wires of our mobile phone are "pushed" by the chemical reaction in the battery. The chemical reaction produces a voltage, which is equivalent to the pressure in our flowing water analogy. The voltage is the energy expended in driving each coulomb of electrons (about six billion billion electrons) through the circuit. A battery producing one volt expends one Joule of energy for each coulomb of electrons pushed through the circuit.

Many different types of batteries have been developed, based on different chemical reactions, and several types are in common use. If we wanted to invent our own battery, we would choose a chemical reaction in which one material is oxidised (that is, loses electrons), and another material is reduced (gains electrons). We should chose a reaction that releases a large amount of energy, since we want to get the maximum electrical energy from a battery of minimum size and weight. Furthermore, it is desirable that the materials to be oxidised and reduced should have low atomic or molecular weights, to provide the highest energy output for the least battery weight.

Normally, in a chemical reaction, the reactants are mixed together so that electrons are transferred directly from one atom to another. In this way, the reaction occurs in an uncontrolled way. Once the reaction starts, heat is released, and higher temperatures cause the reaction to continue at a faster and faster rate until the reactants are consumed.

To make a useful battery, we don't want energy released by the reaction to be released as heat. We want to tap the energy directly to do useful work in pushing electrons through our mobile phones or other devices.

Inside a battery, we want to keep the oxidised and reduced material separated from each other, so electrons cannot transfer directly from one atom to the other. The only way for electrons to be transferred, and for the reaction to occur, is to flow through an external circuit (like our mobile phone). The chemical reaction should only occur when we turn on a switch, allowing electrical current to flow.

One type of battery using relatively simple chemistry is the sodium-sulfur battery. Although this type of battery is not widely used at present, it provides a good illustration of the principles involved. Sodium, being a Group 1 metal, has a strong tendency to lose an electron, while sulfur (a Group 6 element) readily gains two electrons. The reaction produces a relatively large amount of energy – enough to produce very high temperatures when sodium is allowed to react directly with sulfur.

A short (30 second) video shows the reaction when sodium and sulfur are heated in a test tube:
https://www.youtube.com/watch?v=GkcIdn1d_7Y
(note: I is capital letter "I", and 1 is number "1")

A similar reaction, the reaction of zinc with sulfur, is shown in the 3-minute video:
https://www.youtube.com/watch?v=IqLlmXmMiFQ
(note: l is lower case letter "L")

The chemical reaction in a sodium-sulfur battery is:

$$2Na + S \rightarrow Na_2S + \text{Energy (400,000 Joules)}$$
Sodium   Sulfur   Sodium sulfide

The sodium-sulfur battery operates at temperatures of 300 - 350°C, which is above the melting point of sodium and sulfur. Liquid sodium is held in a steel container, which comprises the negative electrode. When an electric current is drawn from the battery, sodium atoms lose electrons, which flow through the "load", and then into the positive electrode, where the electrons combine with sulfur atoms to form sulfide ions.

The key to the battery is a special ceramic material (beta alumina) that separates the liquid sodium and liquid sulfur. The beta alumina ceramic serves two critical functions:

It blocks the passage of sodium and sulfur atoms, which is very important (since if hot liquid sodium and sulfur mixed, they would explode).

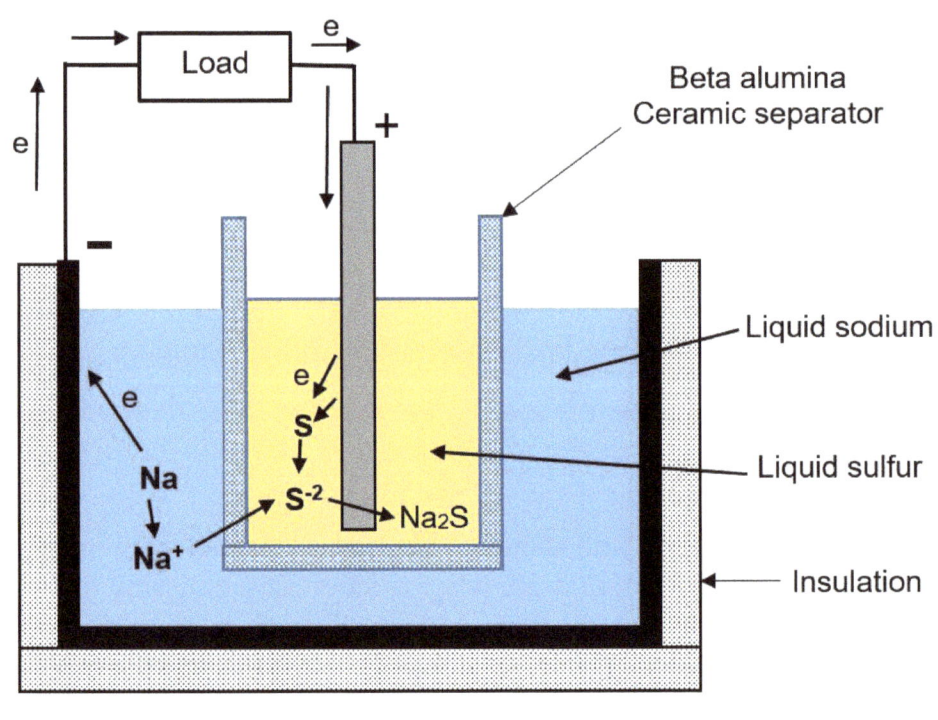

At the high temperatures at which the battery operates, the beta alumina ceramic allows the passage of sodium ions. This completes the circuit for electrical charge to flow through the battery and the load, and allows sodium and sulfide ions to combine.

Sodium ions (Na^+) that form at the negative electrode migrate through the ceramic separator and combine with sulfur ions (S^{-2}) to form sodium sulfide (Na_2S). The reactions occurring at each electrode are as follows:

Negative electrode: $2Na - 2e \rightarrow 2Na^+$
Positive electrode: $S + 2e \rightarrow S^{-2}$
Overall reaction: $2Na + S \rightarrow Na_2S$

The reaction of two moles of sodium and one mole of sulfur produces 400,000 Joules of energy. This energy is expended on driving two moles of electrons through the load, so

200,000 Joules of energy are expended per mole of electrons that flow through the circuit.

One mole of electrons comprises 96,500 coulombs, so the voltage produced by the cell (that is, the number of Joules of energy driving each coulomb of electrons through the circuit) is:

$$\frac{200,000 \text{ Joules/mole of electrons}}{96,500 \text{ coulombs/mole of electrons}} = \text{approximately 2 Joules/coulomb} = 2 \text{ volts}$$

Thus, the sodium-sulfur battery can produce a voltage of about 2 volts.

The balanced equation for the reaction also allows us to work out the maximum energy density of a sodium-sulfur battery. Imagine that we construct a sodium-sulfur battery containing two moles of sodium and one mole of sulfur. Let's, for now, ignore the weight of the steel electrodes, ceramic separator and insulation, and assume that all the energy released by the reaction is converted to electrical energy.

Let's work out the total mass of sodium and sulfur required to construct the battery.
- The atomic weight of sodium is 23, so two moles of sodium has a mass of 2 X 23 grams = 46 grams.
- The atomic weight of sulfur is 32, so the one mole of sulfur contained in the battery would have a mass of 32 grams.

Thus, the total mass of sodium and sulfur is 46 + 32 = 78 grams, or 0.078 kilograms.

The total energy released by reacting two moles of sodium with one mole of sulfur is 400,000 Joules (0.4 megajoules), so the theoretical energy density of the battery is:

$$\text{Theoretical energy density} = \frac{0.4 \text{ Megajoules}}{0.078 \text{ kilograms}} = 5.1 \text{ Megajoules/kilogram}$$

Usually, energy density is expressed in terms of watt-hours per kilogram. Since 3.6 megajoules is equal to one kilowatt-hour (1,000 watt-hours), the theoretical energy density is 1,400 watt-hours per kilogram. This is an extraordinarily high energy density – much greater than is actually achieved in practice. The real energy density is much lower because we have only considered the mass of sodium and sulfur in the battery. The steel shell of the battery, the beta alumina separator, insulation and other components add additional weight, which reduces the energy density. The actual energy density depends upon the construction of the battery, and is reported to be up to 200 watt-hours per kilogram. This compares very favourably with other battery types.

The reaction is reversible, so sodium-sulfur batteries can be charged and discharged thousands of times, making them suitable for energy storage applications.

The sodium-sulfur battery has some very desirable characteristics, and was once considered as a strong contender for electric vehicles. As mentioned, sodium-sulfur batteries have a high energy density (and so, can provide a large amount of energy per kilogram). They have a moderately high power density, so a relatively small battery can meet high power demands, such as when a car accelerates or climbs a steep hill. Sodium-sulfur batteries are also very efficient (providing about 90% of the energy used to charge it), and have long cycle lifetime (can be charged and discharged thousands of times). The chemical reagents used in the battery, sodium and sulfur, are cheap and widely available.

However, sodium-sulfur batteries have several important limitations:

- Sodium-sulfur batteries only operate at temperatures above 300°C. Maintaining such high temperatures is only practical for large batteries, so this effectively precludes their use in small portable electronic devices.

- The two reactants, sodium and sulfur, present a very high risk of fire - particularly at the very high temperatures at which the battery operates:
    - Sodium is highly chemically reactive. It reacts spontaneously with air or moisture and, of course, hot molten sodium would burn or explode if it came into contact with molten sulfur (or with water!).
    - Sulfur will also burn, to produce highly poisonous sulfur oxide fumes.

    Because of these potential hazards, the use of sodium-sulfur batteries has been largely ruled out for electric cars. Use of sodium-sulfur batteries has been limited to energy storage by electrical utilities – and in even this application, adoption has been set back by a fire at a Japanese power utility. Further research is underway to develop sodium-sulfur batteries that operate at lower temperature, and thus, should be less hazardous.

- Because the hot sodium and sulfur are chemically corrosive, the beta alumina separator has a limited lifetime, so the battery must be replaced or refurbished after several years (regardless of how often it has been charged and discharged).

Our consideration of the sodium-sulfur battery illustrates a general point. Each type of battery technology offers its own advantages and disadvantages, and is best suited for specific types of applications. However, the principles are the same. Each type of battery technology is based on a particular oxidation-reduction reaction, and the energy released by the reaction is used to push electrons through an external circuit. The energy expended on each coulomb of electrons determines the battery voltage, and the energy released by each kilogram of reactants determines the theoretical energy density.

Battery voltage and energy density are two very important characteristics, but other factors also impact on the practicality and effectiveness of a battery for each application. These factors include: the power density (maximum power per unit weight), volumetric energy density (energy stored per unit volume), cost and availability of the reactants and other materials, charge-discharge efficiency, ease of fabrication, toxicity, safety considerations, cycle lifetime and shelf life.

14. More on batteries and fuel cells

The lead-acid battery is one of the oldest battery types, and it is still used extensively. Lead-acid batteries have been used in cars for the past 100 years (ever since cars had starter motors, and needed batteries), and are still found under the bonnet of nearly all cars, trucks and buses. It is the most common battery system used to store electricity in off-grid solar and wind power systems, and has long been used by telephone companies as back-up during power failure. All major businesses and commercial buildings have Uninterruptible Power Systems (UPS) to maintain electric power for their computers in the event of a power outage, and the vast majority of UPS systems are based on lead-acid batteries.

Lead-acid batteries have long been used to power submarines underwater. Since before the first World War, lead-acid batteries have been integral to underwater propulsion and operation of (non-nuclear) submarines in all navies of the world. Even today, non-nuclear submarines use lead-acid batteries to operate below snorkel depth. Each of Australia's Collins Class submarines carries 400 tonnes of lead-acid batteries, providing the capacity to travel submerged up to 850 kilometres. These batteries comprise 13% of the weight of each Collins Class submarine.

Because of its long history and continuing use, the lead-acid battery has become the benchmark against which other battery types are judged. Its chemistry is simple, and the characteristics of lead-acid batteries are very well understood. The energy density is not nearly as high as more recent battery technologies (nickel metal hydride and lithium batteries, for example), but lead-acid batteries can produce high power for a short time (necessary for starting car engines), and the volumetric energy density (energy stored per litre volume) is fairly high. The materials used in lead-acid batteries are cheap and readily available. Because lead-acid batteries have been in use for so long, and are used so widely, supply and recycling networks are well established. You could probably buy a replacement battery for your car in any town or city in even the most undeveloped countries of the world.

The lead-acid battery has electrodes made from lead, one of the "transition metals". Previously, we have discussed the Group 1 and 2 metals in the periodic table, which have respectively one or two electrons in their outer shells. *Group 1* metals (the "alkali metals", like lithium, sodium and potassium) have a strong tendency to lose *one* electron, and consequently, to form positively-charged ions with a charge of +1 (Li^+, Na^+, K^+). Similarly, the *Group 2* metals tend to lose *two* electrons, and to form positively charged ions with a charge of +2 (Be^{+2}, Mg^{+2}, Ca^{+2}). We say that the Group 1 metals can have an oxidation state of +1, and Group 2 metals can have an oxidation state of +2. Pretty simple, eh? But transition metals (like iron, copper, manganese, nickel, and lead) can have two or more oxidation states in different chemical compounds. Lead atoms can lose two electrons and form Pb^{+2} (with an oxidation state +2), or be further oxidised to Pb^{+4}.

For a lead-acid battery in the charged state, the negative electrode is lead metal (Pb). The positive electrode is lead dioxide (PbO_2, where lead has an oxidation state of +4). Both electrodes are immersed in an electrolyte solution of sulfuric acid, H_2SO_4. Sulfuric acid is a strong acid, and is completely ionised into hydrogen ions (H^+) and sulfate ions (SO_4^{-2}).

When the battery is connected to a "load" (say, the starter motor of a car), lead atoms at the negative electrode lose two electrons to form lead ions, Pb^{+2}. However, the lead electrode is immersed in a solution of sulfuric acid, so as soon as lead ions form, they combine with sulfate ions in the electrolyte solution to form lead sulfate ($PbSO_4$). Lead sulfate is a solid. It doesn't dissolve in the electrolyte and remains on the surface of the electrode.

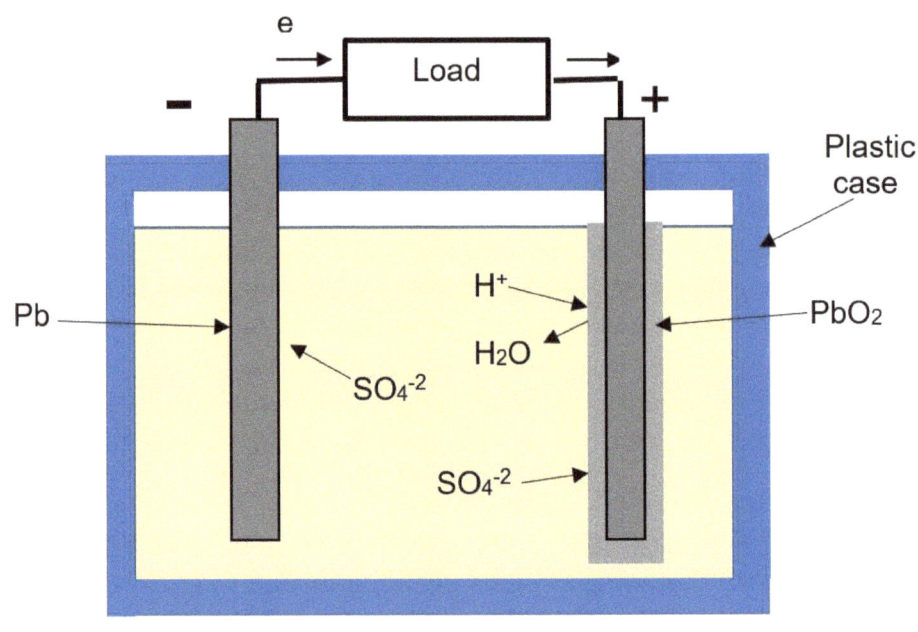

Lead-acid battery cell

The reactions occurring at the negative electrode are:

$$Pb - 2e \rightarrow Pb^{+2}$$
$$Pb^{+2} + SO_4^{-2} \rightarrow PbSO_4$$

Adding these two reactions together gives us the overall reaction for the negative electrode:

<u>Negative electrode:</u> $Pb + SO_4^{-2} - 2e \rightarrow PbSO_4$

Since the loss of electrons is oxidation, lead is oxidised at the negative electrode.

At the positive electrode, Pb^{+4} ions in the form of lead dioxide are reduced. Each Pb^{+4} ion gains two electrons, forming a Pb^{+2} ion. Once again, as soon as Pb^{+2} forms, it combines with sulfate ions in solution to form lead sulfate. Don't forget that the active reactant at the positive electrode is lead dioxide, PbO_2. So you might be wondering: what happens to the oxygen atoms? They combine with hydrogen ions in the electrolyte solution to form water.

The reactions occurring at the positive electrode are:

$$PbO_2 + 4H^+ + 2e \rightarrow Pb^{+2} + 2H_2O$$
$$Pb^{+2} + SO_4^{-2} \rightarrow PbSO_4$$

Adding these two reactions together gives us the overall reaction for the positive electrode:

<u>Positive electrode:</u> $PbO_2 + 4H^+ + SO_4^{-2} + 2e \rightarrow PbSO_4 + 2H_2O$

Adding the reactions occurring at the negative and positive electrodes gives the overall reaction for the lead-acid battery cell:

<u>Overall reaction:</u> $Pb + PbO_2 + 4H^+ + 2SO_4^{-2} \rightarrow 2PbSO_4 + 2H_2O + Energy$

As the battery is discharged, both electrodes are converted to lead sulfate ($PbSO_4$), and the sulfuric acid electrolyte solution becomes increasingly dilute.

The maximum power output of a lead-acid battery reduces at lower temperatures – as it does for all types of chemical batteries. The decline is particularly acute at temperatures below freezing. This can present serious problems for those who drive cars in very cold winter climates, as I did as a young adult living in New York City, driving my father's badly-rusted and poorly maintained 1958 Oldsmobile. Cars of this generation did not have electronic ignition or fuel injection, as they do now, so the engine often would not start straight away if the fuel/air mixture wasn't quite right. On a really frigid winter morning, if the engine didn't start straight away, it often wouldn't start. At sub-freezing temperatures, not only does the car battery have greatly reduced power output, but the engine oil becomes thick and gluggy, so even more power is needed to crank the engine.

I learned that sitting in a car that won't start is terribly frustrating. I have seen drivers persist in trying to start the engine, even though the starter motor hardly rotated the engine (the term "flogging a dead horse" comes to mind). However, as repeated attempts are made to start the engine, the discharge reaction within the battery converts the sulfuric acid electrolyte to water. If the battery is left in a discharged state, the electrolyte might then freeze, expand and crack the battery case (with dilute sulfuric acid leaking out and corroding the surrounding metal).

During my early years of driving experience, I observed that the first really cold day of winter would expose any car battery which was approaching the end of its life. Starting the car on this day provided a simple pass-fail test of battery condition.

I have not had to worry about this problem for the past 30 years, while I have been living in Brisbane, where winter temperatures hardly ever go below freezing. Based on my experience during this time, I have concluded that car batteries in sub-tropical Australia have a completely different mode of failure at the end of their life. Here, there is no "winter starting test" to screen out weak batteries, so it seems that when batteries reach the end of their life, they just completely die and stop working.

But, I have digressed! Let me return to the discharge reaction of the lead-acid battery. The energy released by the reaction drives electrons through the circuit with a cell voltage of about 2 volts (the common 12 volt battery contains six cells connected in series). The reaction of one mole of lead with one mole of lead dioxide produces about 400,000 joules of energy (about the same as is produced in a sodium-sulfur battery by the reaction of one mole of sodium and one mole of sulfur). However, the reactants for the lead-acid battery (particularly, the lead and lead dioxide electrodes) have high atomic and molecular weights. One mole of lead, with a mass of about 207 grams, weighs nearly ten times as much as a mole of sodium. One mole of lead dioxide, PbO_2, has a mass of 207 + 2 X 16 = 239; and two moles of H_2SO_4 has a mass of 196 grams. Consequently, the reactants to produce 400,000 Joules of energy would have a total mass of 642 grams (0.642 kilograms). Of course, the plastic case, water and other components also have significant mass, but let's ignore those for now. The maximum possible energy density of a lead-acid battery is

$$\text{Maximum energy density} = \frac{0.4 \text{ Megajoules}}{0.642 \text{ kilograms}} = 0.62 \text{ Megajoules/kilogram}$$

Expressing the maximum energy density of a lead-acid battery in terms of watt-hours per kilogram, we find that the maximum theoretical energy density is 170 watt-hours per kilogram (about one-eighth that of a sodium-sulfur battery). In fact, the actual energy density of a lead-acid battery is usually considered to be about 60 watt-hours per kilogram.

One of the particular characteristics of a lead-acid battery is that its lifetime is dramatically reduced if it is fully discharged. The number of charge-discharge cycles that a lead-acid battery can provide in its lifetime is strongly affected by the "depth of discharge". In normal operation, a lead-acid battery should never be discharged below half capacity. So, the actual *usable* energy storage is much less than the maximum energy density allowed by the reaction.

Fuel cells

A fuel cell is an electrochemical cell that oxidises a fuel (usually hydrogen or natural gas), with the combustion energy of the fuel converted directly into electricity. Although a fraction of the combustion energy is released as heat (due to inefficiencies within the fuel cell), fuel cells are *not* heat engines, and are *not* subject to the efficiency limitations which apply to heat engines. Fuel cells are fundamentally different from "normal" engines or electricity generating plants, which convert fuel energy into heat, which is then converted into mechanical power.

The most efficient fuel cells can now achieve efficiencies of 50-60% (that is, convert 50-60% of fuel energy into electricity). In principle, fuel cells could ultimately achieve efficiencies up to 100%. This contrasts with the best heat engines, which are about 35% efficient.

Various types of fuel cells have been developed. Some operate at relatively low temperature (about 60-80°C), and these have attracted the most interest for fuel cell vehicles. Most prominent among these are the PEM (Polymer Electrolyte Membrane) fuel cells.

Low temperature fuel cells are generally limited to using hydrogen as fuel. To get the electrode reactions to occur sufficiently quickly, low temperature fuel cells incorporate rare and expensive "noble metals" (such as platinum and palladium) catalysts in the electrodes. To avoid "poisoning" the catalysts, the hydrogen fuel must free of impurities (such as carbon monoxide and hydrogen sulfide).

PEM fuel cells use a special polymer membrane as a solid electrolyte. This membrane prevents the passage of hydrogen and oxygen gas between electrodes, but allows hydrogen ions to pass. This ion-conducting membrane allows electrical charge to

PEM Fuel Cell

Diagram of PEM fuel cell showing: Porous negative electrode with H_2 input, Polymer Electrolyte Membrane in the middle with H^+ ions passing through, Porous positive electrode with O_2 input, and H_2O (steam) output. Load connected across electrodes with electron flow (e) from negative (-) to positive (+) terminal.

Reactions:
- Negative electrode: $2H_2 - 4e \rightarrow 4H^+$
- Positive electrode: $O_2 + 4H^+ + 2e \rightarrow 2H_2O$

complete its passage from one electrode through the load to the other electrode, and then within the cell.

Pure hydrogen gas is passed over a porous negative electrode. A catalyst on the electrode surface breaks hydrogen molecules into atoms, which can then lose an electron to the electrode. Electrons pass through the load to the positive electrode, which is also porous and contains a catalyst. Oxygen (air) is passed over the positive electrode, is broken into oxygen atoms, and then gains electrons to form O^{-2} ions. These combine with H^+ hydrogen ions passing through the polymer membrane to form water. Water vapour (steam) is the only product of the reaction.

The reactions occurring at the negative and positive electrodes are as follows:

$$\text{Negative electrode:} \quad 2H_2 - 4e \rightarrow 4H^+$$
$$\text{Positive electrode:} \quad O_2 + 4H^+ + 4e \rightarrow 2H_2O$$

Adding the reactions at the negative and positive electrodes gives the overall reaction of the fuel cell:

$$\text{Overall reaction:} \quad 2H_2 + O_2 \rightarrow 2H_2O \text{ (steam)} + 458{,}000 \text{ Joules}$$

This is the same reaction that occurs when hydrogen burns in air. Oxidation of two moles of hydrogen (as shown in the reaction above), releases 458,000 Joules and involves the transfer of four moles of electrons. This corresponds to 1.2 Joules of energy[Note 1] expended in pushing each coulomb of electrons through the load. In other words, the cell can produce a maximum voltage of 1.2 volts. Actual cell voltage is usually about one volt or slightly less. Many fuel cell modules are stacked in series to produce several hundred volts. Higher voltages are desirable for most applications, including operating electric motors to propel electric vehicles.

Fuel cells are ideal for powering cars in congested inner cities because they produce no pollution. Hydrogen fuel is oxidised with air, producing water vapour as the only product. There is no need to fit anti-pollution equipment to get rid of nitrogen oxides, unburned hydrocarbons and carbon monoxide pollutants that are produced by internal combustion engines (which currently power our cars, trucks, buses, lawnmowers and construction vehicles). Also, fuel cells have few or no moving parts and are extremely reliable. They are virtually silent. The main sound made by fuel cell vehicles is the whine of an electric motor.

However, the requirement to use hydrogen fuel is problematic. Although hydrogen is, by far, the most common element in the known universe, it does not occur naturally within the Earth's crust (as does, say, natural gas, petroleum and coal). Hydrogen must be produced from other energy sources, as was discussed in the chapter on "The transport revolution".

Another problem is that hydrogen is difficult to store, especially on board vehicles, where weight and space available for fuel storage are very limited:

- Hydrogen can be stored as a high-pressure gas, but this requires large and heavy storage tanks made of steel or carbon fibre composites. However, even at very high pressure (700 atmospheres), the energy storage density is only one-sixth as much as gasoline, diesel fuel or jet fuel.

- Hydrogen can be stored as a cryogenic liquid (at about -252°C). Liquid hydrogen has about one-quarter the energy per unit volume of gasoline, and must be stored in special insulated storage tanks. Even with the best insulation, liquid hydrogen will gradually boil off. This is not a problem if the evolved hydrogen gas can be used immediately, but if a vehicle is not used for long periods, the evolved hydrogen gas must be released and wasted.

- Hydrogen can be adsorbed onto magnesium or other metal alloys. This allows hydrogen to be stored at much lower pressure than as compressed gas, but releasing the absorbed hydrogen at the desired rate poses technical challenges.
- Hydrogen could be produced on-board a vehicle by, say, gasifying diesel fuel, but this requires complex equipment, with attendant problems of cost, reliability, and additional weight and space within the vehicle.

Perhaps the biggest barrier to the adoption of hydrogen fuel cell cars is the classic "chicken and egg" conundrum that faces many new technologies. Vehicle manufacturers cannot produce and sell hydrogen-powered vehicles until fuel production and supply infrastructure is in place. Who would be willing to buy a hydrogen-powered vehicle if there are hardly any refuelling station in their city and along main highways? But which companies would be prepared to invest in hydrogen production facilities, distribution pipelines and refuelling stations if there are very few hydrogen-powered vehicles to refuel?

But these problems could be overcome. A limited number of fuel cell vehicles have been produced in the past by Toyota and Mercedes Benz. It appears that Toyota is beginning production of the first mass-produced fuel cell vehicle in 2015. Presumably, the company will focus its sales efforts on specific areas where hydrogen refuelling infrastructure is being established. A website with impressive graphics of this Mirai fuel cell vehicle can be viewed at:

http://www.toyota.com/fuelcell/fcv.html

Notes:

(1) Energy/coulomb = $\dfrac{458{,}000 \text{ Joules}}{(4 \text{ moles of electrons})(96{,}500 \text{ coulombs per mole of electrons})}$

= 1.2 Joules/coulomb = 1.2 volts

15. Fuel cell, electric and hybrid vehicles

Transport of people and goods is a fundamental part of our economy and society. Cars, buses, trucks, aircraft, trains, ships, construction vehicles and farm vehicles rely almost entirely on fuels derived from crude oil. Conventional vehicle technology is based on reciprocating internal combustion engines – namely gasoline engines (which operate on spark ignition) and diesel engines (operating with compression ignition). We have already considered the shortcoming of these engines in terms of the air pollution that they release, and their reliance on crude oil (of which many countries, including Australia, are likely to become increasingly reliant on imports).

Energy, and work, must be expended in pushing cars and other vehicles along a road. This requires an engine or other source of mechanical power to drive the vehicle along the road:

- For one thing, the drive system must provide the force required to overcome the rolling resistance of the vehicle's tyres on the road. Rolling resistance varies directly with the weight of the vehicle, as well as the properties of the tyre and road surface.

- The drive system must overcome the drag force of air resistance to push the vehicle through the air. Air resistance varies dramatically with vehicle speed. In fact, it varies with the velocity to the third power (so eight times as much force is required to push the vehicle twice as fast). The air resistance does not depend on the weight of the vehicle, but varies with size and shape of the vehicle (how streamlined it is).

- Energy is required to accelerate the vehicle to highway speeds – that is, to impart kinetic energy to the vehicle. This is an initial energy investment. Once the vehicle reaches the desired speed, further energy is only required to overcome rolling resistance and air resistance to maintain this speed. However, in real driving conditions, vehicles must often slow down or stop for traffic lights, stop signs or traffic obstructions. When the driver presses the brake pedal, kinetic energy that was imparted to the vehicle is dissipated and lost as waste heat. For electric vehicles, there is potential for "regenerative braking", which converts kinetic energy of the vehicle into electrical energy which is stored in a battery (and may be used to accelerate the vehicle once again).

- Energy must be expended to push a vehicle uphill. This energy is invested in increasing the gravitational potential energy in raising the vehicle to higher elevation. The gravitational potential energy could – in principle - be recovered when the vehicle travels downhill. After all, on any given round trip, the elevation gained in going uphill is exactly equal to the elevation lost in going downhill (since the vehicle starts and finishes at the same elevation). However, it is often not practical to recover all the energy invested in pushing a vehicle uphill. For example, the gravitational potential energy gained when a one-tonne car (with a weight of 10,000 Newtons) is driven up the Toowoomba range (with an elevation of around 500 metres) is 10,000 X 500 = 5 million Joules. Most likely, when the vehicle is driven down the range, the driver will need to ride the brakes to keep the vehicle at safe speed, dissipating and wasting most of the energy invested when the vehicle drove up the range.

- Energy is required to operate various auxiliary functions, such as air-conditioning, power steering, headlights, rear window defroster, power brakes and a myriad of other features and gizmos which are often included in modern cars (electric windows, seat adjustment, mirrors adjustment, tailgates, central door locking, etc).

Conventional vehicles use gasoline or diesel engines as a power source to meet these energy requirements. Most cars in Australia and the United States have gasoline engines (diesel engines are much more common in European cars). A typical gasoline engine in a car is **nominally** about 30% efficient (that is, converts 30% of the energy contained in the fuel into mechanical power). However, this nominal efficiency is only achieved when the engine is operated under optimal conditions at its optimal rated load. However, in real vehicles, driven on real roads, the engine is rarely operated at its optimal rated load. Under actual driving conditions, a car is driven at various speeds (including idling at a standstill), and frequently accelerates and then slows down. The engine must be sufficiently large and powerful to provide the **maximum power requirements** expected by the driver (to pass vehicles on a highway; accelerate at traffic lights; drive up steep hills; or to hoon, burn rubber and be an obnoxious hazard to other road users). This means that, 95% of the time, the engine will be much larger than is necessary (typically, around 4 times larger than required to maintain a vehicle at highway speed).

One obvious consequence is that the vehicle engine is heavier and takes up more space than what is needed 95% of the time. Another consequence, which may not be obvious, is that the engine will operate at sub-optimal efficiency for most of the time. The efficiency of a gasoline engine (and this applies also to diesel engines and gas turbines) depends on the "load" (how much power the engine is required to produce).

If we were to plot a graph of the efficiency of a gasoline engine, versus its power output, it would look something like this:

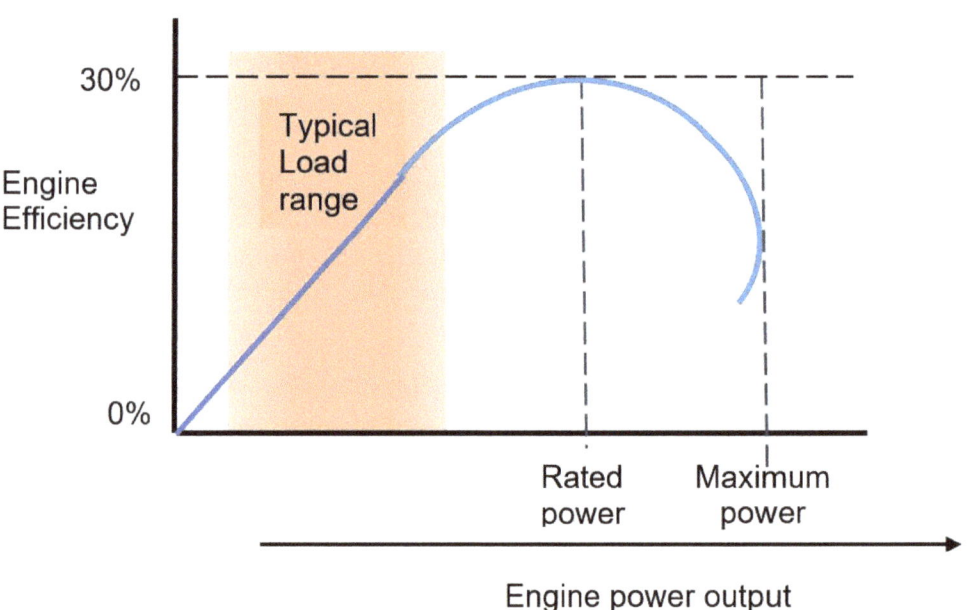

When the engine is idling (say, when the car is at a traffic light or sitting in heavy traffic), it is not producing any useful work – but yet, the engine is still consuming significant amounts of fuel. At idle, there is zero load (no power is required or produced), and the engine is 0% efficient. As the load on the engine increases (and as it produces more power), its efficiency increases until it reaches maximum torque and maximum efficiency. This is the "sweet spot" at which the engine is most efficient, but rarely does the engine need to produce this much power. Rather, in typical driving conditions, the engine spends most time producing a fraction of its rated output. So, while the engine might be capable of achieving an efficiency of 30%, its actual efficiency would probably be half that under typical driving conditions.

The concept of a "hybrid vehicle" originally emerged as a means to operate the engine within a load range that closely matches its optimal rated power output. The idea is to incorporate a much smaller engine, one whose rated power output is just enough to keep the car moving at highway speed. The vehicle would also include an electric drive system (a battery and electric motor) to augment the power output of the gasoline engine when high power is needed for acceleration or hill climbing. During periods of low-load, the gasoline engine would have surplus capacity and could charge the battery. The battery and electric motor would "even out" the load on the engine, so that it produces relatively constant power output at its optimal operating condition.

Schematic diagram of a "parallel hybrid vehicle"

When little power required, small engine drives car and recharges battery. Engine operates near optimal load.

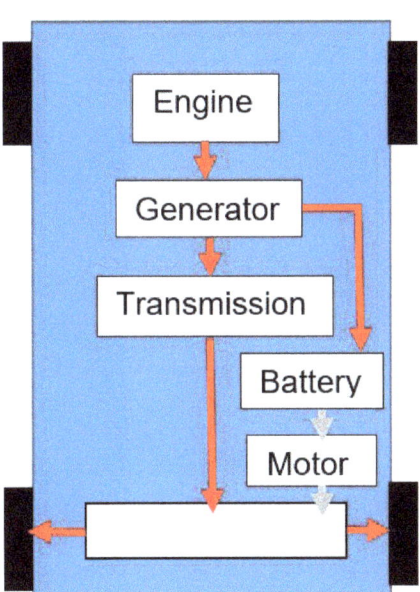

When high power required, engine power is augmented by electric motor.

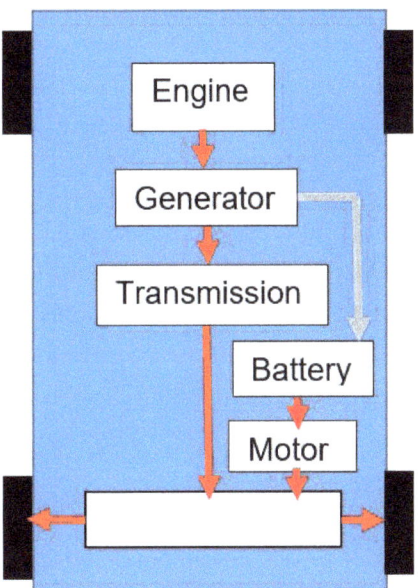

The problem with this concept is that it required two propulsion systems (a gasoline engine and electric drive), and this adds to the cost and weight of the vehicle. Particularly, in the 1970s and 1980s, the available lead-acid battery technology was not up to the task. To provide the required energy storage capacity and high power output, the battery would weigh several hundred kilograms, and would significantly increase the total weight of the car. The additional weight due to the battery would require additional power for acceleration and hill climbing, as well as heavier suspension and brakes – requiring an even bigger and heavier battery to achieve the desired acceleration and hill climbing. Hybrid vehicle design became a "cat chasing its tail" exercise.

Development of superior battery technology (such as nickel metal hydride and lithium batteries) in 1990s and 2000s opened the way for practical hybrid vehicles. Toyota introduced the Prius hybrid, and other car manufacturers have followed suit. However, these vehicles differ from the original concept of a hybrid vehicle envisaged decades ago. These vehicles are "light hybrids". The gasoline engine is only slightly smaller than in a conventional car, and the battery/motor system has relatively small energy storage and power output.

In such "light hybrids", the main function of the electric drive is **not** to *augment* the engine when maximum power is required for passing or climbing steep hills, but rather, to enable the engine to be shut off when only small amounts of power are required (in the load range at which the gasoline engine is least efficient). The on-board computer shuts off the gasoline engine when the vehicle is stopped at a traffic light (if the battery is charged and the engine is at operating temperature). While the engine is off, the electric drive system operates the air-conditioning, headlights and other accessories. So long as low power is required - while the vehicle is driving at low speed in heavy traffic or while parking - the vehicle is powered solely by the battery drive. However, once the driver presses on the accelerator, the computer starts and engages the gasoline engine to meet the larger power requirements (and recharge the battery). The electric drive and computer system ensure that the engine rarely operates in the inefficient low-load range.

As battery technology continues to improve, car makers are likely to introduce hybrid vehicles with greater energy storage and more powerful electric drive motors (conforming to vehicle design strategy originally envisaged for hybrid vehicles). Provision may be made for the batteries to be recharged from the power grid while the vehicle is parked, as well as by the on-board engine.

At the same time, car manufacturers may follow the all-electric route, and produce vehicles powered solely by batteries and electric motors. To provide for extended range of electric vehicles, some companies are proposing that electric recharging stations be located at carparks in shopping centres and workplaces, while other companies are developing systems that would replace the entire battery pack (with a fully charged unit) from underneath the vehicle using a fully-automated winching system. A Brisbane-based electronics company, Tritium, has developed fast-charging systems that can recharge an electric vehicle battery in about 20 minutes. This is probably the most advanced fast-charge battery technology in the world. (See: http://tritium.com.au/products/veefil).

As an alternative strategy, vehicle manufacturers may try to overcome the efficiency limitations of internal combustion engines, and their reliance on crude oil, by using fuel cells. Most major manufacturers have developed prototype "fuel cell concept vehicles". The first major applications of fuel cell vehicles will likely be in fleets with central refuelling facilities, as this would overcome the initial lack of refuelling infrastructure. Trials of fuel cell-powered buses have been held in various cities of the world, including in Perth, Western Australia. These trials demonstrated that fuel cell buses could meet all requirements of local councils (the bus operators) and commuters. The buses provided the same performance as the diesel buses that they replaced, except for being much quieter and producing no fumes or smoke.

Take a one-minute ride on a fuel cell-powered Mercedes Benz bus in Perth trial, 2004
https://www.youtube.com/watch?v=skpgmaQgzR0

More recently, hydrogen-powered fuel cell buses operated at the 2010 Olympics in British Columbia, Canada. Here is a link to a 2-minute video:
http://www.centreforenergy.com/AboutEnergy/FuelCell/Videos.asp

In recent years, we have started to see electric-powered and electric-assisted bicycles, and a really cute fuel cell bicycle has recently been developed by a university in Sydney. Here is a 1-1/2 minute video of the fuel cell bicycle. It produces its own hydrogen by using electricity from the grid to electrolyse water. Although it uses a fuel cell, it is essentially an electric vehicle, using a hydrogen tank for on-board energy storage (instead of a battery). It goes

to show that "hybrid vehicles", "electric vehicles" and "fuel cell vehicles" are not necessarily separate and distinct technologies, but overlap and represent a continuum of approaches:
https://www.youtube.com/watch?v=kR5rpVdMtXM

Here is a short video (1-1/2 minutes) about fuel cell vehicles that you might like to look at:
http://auto.howstuffworks.com/fuel-efficiency/4836-how-fuel-cells-work-video.htm

16. Petroleum fuels

Currently, the major sources of energy used for electricity generation, transport, industrial heating and household use are fossil fuels – natural gas, crude oil and coal. Most coal, oil and gas were formed during the Carboniferous period, from 350 to 300 million years ago. Geological forces were active during this time, causing land to subside across large areas of the Earth's surface, forming swamps and shallow seas. These conditions were optimal for the formation of fossil fuels. As the land subsided, plants, algae and other living organisms washed into shallow seas during periods of heavy rainfall, collected on the bottom, and were buried beneath layers of sand and silt. The organic matter was quickly covered by sediment, protecting it from contact with oxygen in the air and from normal decay processes (in which bacteria oxidize carbohydrates, proteins and fats to produce carbon dioxide and water).

Over millions of years, as organic matter collected and became buried under kilometres of sediment, heat and pressure transformed the material into what we now know as natural gas, crude oil and coal.

Plants, algae and other living organisms are composed primarily of the elements carbon, hydrogen and oxygen (with smaller amounts of nitrogen, phosphorus, sulfur and other elements). As organic matter was subjected to heat and pressure deep within the Earth's crust, these chemical elements re-arranged to form hydrocarbons (chemical compounds composed of carbon and hydrogen). Oxygen that was originally present reacted with some of the carbon to form carbon dioxide (or reacted with hydrogen to form water).

Hydrocarbons that were formed were either in the form of gas, liquid or solid. Often, hydrocarbon gases and liquids migrated through porous rock layers until they were eventually trapped beneath layers of impervious rock. There, they remained as deposits of natural gas or crude oil (sometimes both are found together). In some cases, the organic matter formed solid material, which remained in the rock formation as bands or beds of coal. Some tarry organic material collected within the pores of sand or clay formations, forming tar sands or oil shale. Increasingly, these "non-conventional" sources of oil are being mined as a source of fossil fuels.

Fossil fuels are . . . fossilised fuels, remnants of the period when they formed many millions of years ago (before dinosaurs appeared on Earth). Fossil fuels are no longer being formed at a significant rate, so the world is currently relying on a once-off endowment from the distant past. Basically, humanity is living off the inheritance bequeathed by ancestral lifeforms, distantly related to species from which we evolved long, long ago. Fossil fuels formed over hundreds of millions of years, but they are being consumed within a few hundred years. The human species *homo sapiens* has only existed for about 100,000 years, and the modern industrial age is only about two hundred years old. There is virtually no possibility that new deposits of fossil fuels will be formed within the time span of our species.

Natural gas, crude oil, coal and "non-conventional" oil sources are non-renewable resources, and their current usage is clearly not sustainable as the main energy source for human society for subsequent generations. However, the world will not suddenly run out of fossil fuels. Major improvements in technology have enabled resources to be extracted that were previously not considered economically recoverable. As the cheapest and most accessible fuel reserves are depleted, we will increasingly rely on hydrocarbon resources that are deeper, more difficult to extract and located in more remote and undersea locations (with attendant costs and environmental risks), and on non-conventional sources whose recovery and processing entails higher financial and environmental costs.

The chemistry of natural gas

Natural gas consists primarily of methane, the simplest hydrocarbon. The methane molecule has a central carbon atom bonded to four hydrogen atoms. Carbon is a Group IV element with 4 electrons in its outer shell. Since it sits in the middle of the periodic table, it is neither a strong metal nor a non-metal, and always forms four covalent bonds – that is, shares its four outer-shell electrons with other atoms.

The structure of methane is commonly depicted as shown here:

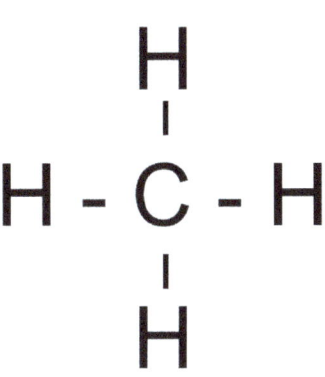

This structure accurately conveys that the carbon atom is located centrally, bound to four surrounding hydrogen atoms. However, such a two-dimensional drawing does not convey the three-dimensional shape of the molecule. The four hydrogen atoms are not within a plane, but located at the corners of a tetrahedral pyramid, with a 109° angle between the axes of the carbon-hydrogen bonds. This means that four hydrogen atoms in methane are equivalent and equally far apart from one another.

The actual 3-dimensional shape of a methane molecule can be better appreciated by a model, shown here. The central axis of each chemical bond is represented by a spring. The central carbon atom is shown in black, and the four hydrogen atoms are white.

If the molecular model is flipped, so that one of the hydrogen atoms forming the base is brought vertically, the resulting shape will be identical to the original.

Natural gas is widely used by households (for cooking and water heating), in industry (metal melting in foundries, cement manufacture and glass production) and also for electricity generation (in gas turbine power stations, especially for peak and intermediate-load generation).

When natural gas is burned, the methane reacts with oxygen in the air. The chemical reaction is:

$$CH_4 + 2\,O_2 \rightarrow CO_2 + 2\,H_2O + \text{Heat}$$

Methane + Oxygen → Carbon dioxide + Water vapour + Heat

This balanced equation tells us that one mole of methane (with a volume of about 25 litres) reacts with two moles of oxygen (50 litres) to produce one mole of carbon dioxide and two moles of water vapour. The reaction is highly "exothermic", with 890,000 Joules of heat produced for each mole of methane that burns. The carbon-oxygen and hydrogen-oxygen bonds formed by the reaction are much stronger (are a lower state of energy) than the chemical bonds within methane and oxygen. The reaction produces heat because the reactants (one mole of methane and two moles of oxygen) have more chemical potential energy than the products, so the reaction "runs downhill".

The release of energy by the oxidation of methane (or burning of any hydrocarbon) is exactly analogous to that experienced by anyone who has ridden a bicycle on a hilly route. At the top of a hill, the cyclist has a high state of gravitational potential energy. As the cyclist rides down the hill, he/she gains speed as the initial gravitational potential energy is converted to kinetic energy. Eventually, the kinetic energy acquired by the bicycle rider is degraded to heat by air resistance and turbulence. The amount of heat produced is equal to the loss of gravitational potential energy.

In the same way, methane and oxygen represent a state of high chemical potential energy (equivalent to being at the top of the hill). When we mix methane and oxygen, they do not react until we provide an ignition source, such as a spark. The ignition source provides enough energy to get the reaction to the top of the "energy hill". It weakens the chemical bonds *within* the methane and oxygen molecules, allowing them to form new, stronger bonds (carbon-oxygen and hydrogen-oxygen bonds). The initial energy input required to get the reacting molecules "to the top of the hill" is referred to as the "activation energy". Of course, once a few molecules react, this increases the temperature of surrounding molecules of methane and oxygen, providing the activation energy for these molecules to react.

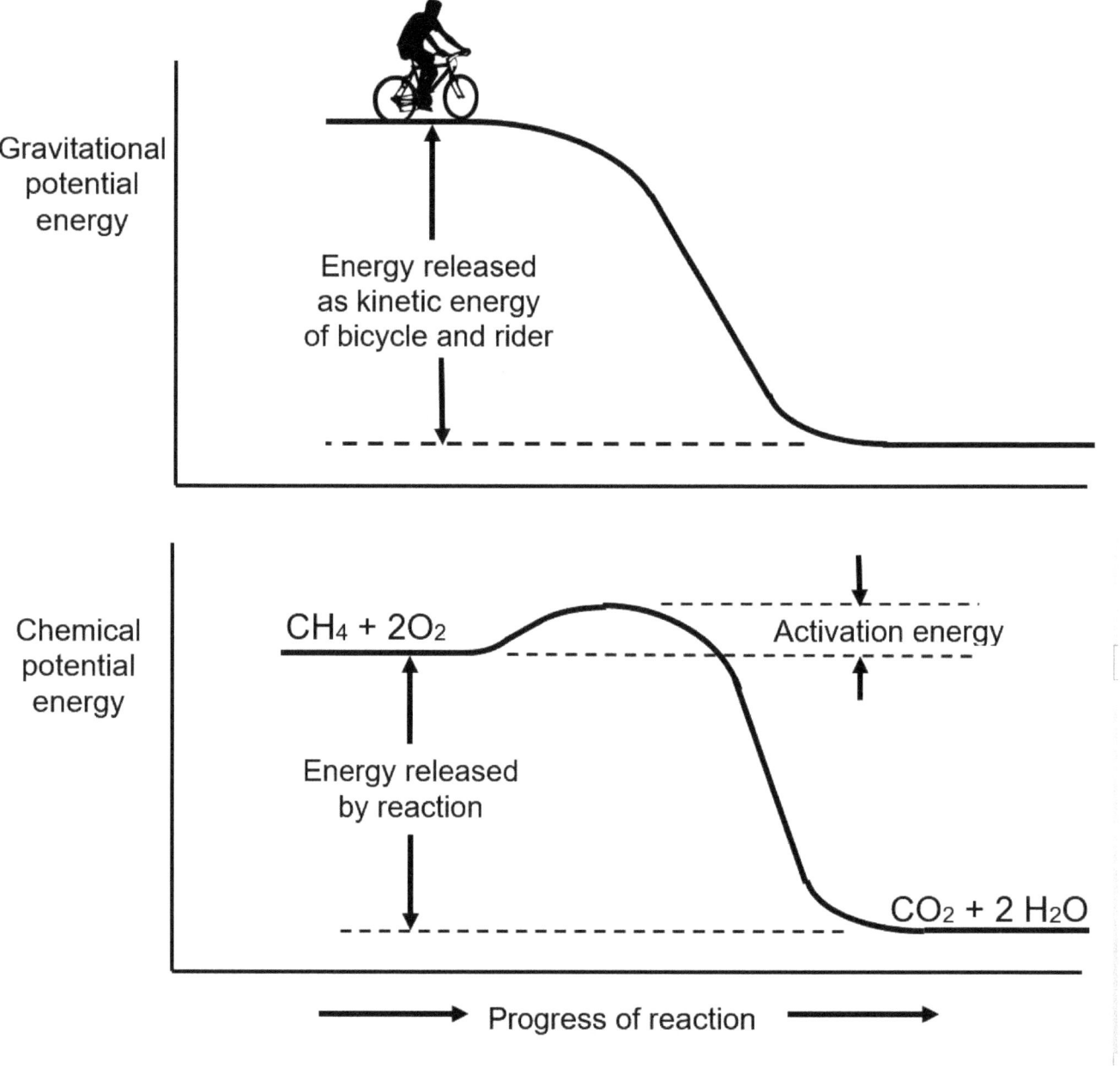

Meet methane's family

Natural gas contains small amounts of other hydrocarbons, which have similar properties to methane. Let's meet methane's "family" of related chemical compounds, called the "alkanes".

Most closely related is the compound **ethane** with two carbon atoms bonded to each other, and to six hydrogen atoms.

Ethane and methane are both relatively small molecules, and have relatively weak attractive forces acting between molecules. As a result, methane and ethane are both gases at normal temperature. Methane must be cooled to -164°C before it liquefies, and ethane has a boiling point of -89°C. Nonetheless, natural gas is increasingly being liquefied and exported from Australia on ships with huge cryogenic storage tanks containing Liquefied Natural Gas (LNG).

For larger alkane molecules – those containing four, five or fifty carbon atoms – the attractive forces between molecules increases with the size of the molecule. These intermolecular forces tend to hold the molecules in the liquid or solid state if the temperature is low enough. Consequently, alkanes with more carbon atoms remain liquid at higher temperatures. They have higher boiling points, and less tendency to vaporise.

Natural gas also contains some **propane** and **butane**, whose molecules contain three and four carbon atoms respectively. Propane and butane are gases at normal air temperature and pressure, but are readily liquefied at modest pressure. Consequently, propane and butane are commonly stored as liquid within pressurised steel tanks, and are referred to as Liquefied Petroleum Gas (LPG). These compounds are separated from natural gas and sold as automotive LPG fuel (which, because it is cheaper than gasoline, is widely used in taxis and other high-mileage vehicles). LPG is also used as fuel for barbecues and camping stoves, and for gas deliveries to households and restaurants without access to natural gas pipelines.

Finally, natural gas contains small quantities of the five-carbon hydrocarbon **pentane** (C_5H_{12}) and the C6 hydrocarbon **hexane** (C_6H_{14}). These compounds are liquid at normal air temperature, but have low boiling points and readily evaporate. Pentane and hexane can easily be separated from natural gas, since they readily condense as the gas is cooled. Pentane and hexane are referred to as "condensate" because they condense into liquid at normal temperatures. Condensate is mixed with other hydrocarbons in gasoline motor fuel. However, because of their low boiling point and high volatility, the amount of pentane and hexane mixed into gasoline must be limited to prevent the fuel from boiling inside fuel lines within the hot engine compartment of cars. In winter climates, more condensate is added to gasoline, as temperatures are lower and more volatility is desirable for cold weather starting.

Crude oil and its products

Alkanes with five or more carbon atoms are found in crude oil. Molecules of these "higher alkanes" contain a chain of carbon atoms, each with a chemical bond to two adjacent carbon atoms and to two hydrogen atoms. Consider, for example, the hexane molecule shown below. Note that each carbon atom *within the chain* is bonded to two hydrogen atoms, while those at each end of the chain (called "terminal" carbon atoms) have an extra hydrogen atom.

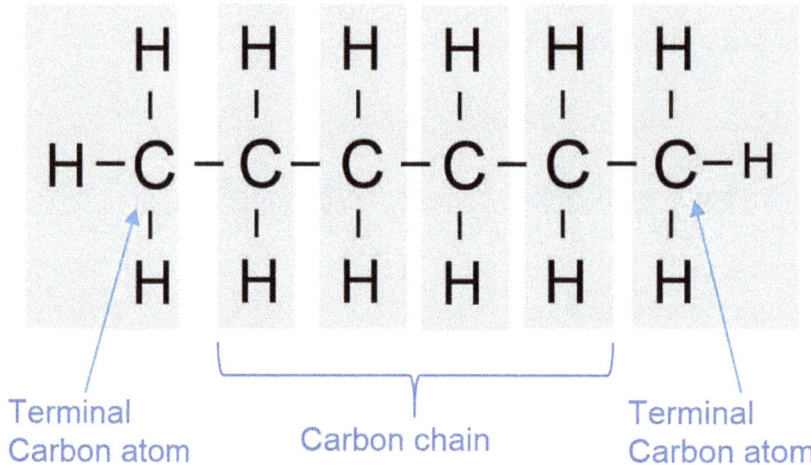

Thus, for any alkane, the number of hydrogen atoms is twice the number of carbon atoms, plus the additional two hydrogen atoms at the ends of the chain. An alkane with **n** carbon atoms will have **2n+2** hydrogen atoms. The chemical formula for any alkane is C_nH_{2n+2}.

Gasoline (also called "petrol" or "motor spirit") is a mixture of hydrocarbons containing from 5 to 12 carbon atoms. This range is chosen to give the desired level of volatility. Gasoline must be easily vaporised to form a fuel-air mixture, as it is fuel vapour that burns inside the cylinders of a gasoline engine.

Different requirements apply to diesel engines, in which the fuel is burned as fine droplets of liquid. Diesel fuel does not need to be volatile, and in fact, the lower volatility of diesel fuel is a major safety advantage. The "flash point" of diesel is above 50°C, so diesel fuel spills will not ignite at normal air temperature, even if an ignition source is present. Diesel fuel contains hydrocarbons ranging from eight to twenty four carbon atoms (C8 to C24).

Jet fuel has a volatility which is in-between that of gasoline and diesel fuel, containing hydrocarbons within the range of eight to eighteen carbon atoms (C8 – C18). This range provides the best trade-off to meet the conflicting requirements for operating jet aircraft. Low volatility is advantageous from a fire safety point of view, but on the other hand, it is also essential that that the fuel doesn't freeze at temperatures around -50°C encountered at high altitude. Commercial jetliners have back-up systems and redundancy to avoid catastrophic problems if any one system should fail, but there is no back-up fuel system. Consequently, jet fuel must meet very strict specifications.

Crude oil contains even larger hydrocarbon molecules that are solid or gooey liquids at normal air temperatures. These hydrocarbons only flow when they are heated, if at all, and are used in lubricating oils and greases, bunker oil (used in large ships), furnace oil and asphalt for roads.

Natural gas and crude oil vary widely in composition. "Raw" natural gas (as it emerges from the ground, and before it is processed) can contain significant amounts of carbon dioxide, which is removed before the gas is compressed and transported by pipeline. The gas can be "wet" or "dry", depending upon the relative amount of condensate. It can contain significant amounts of hydrogen sulfide (which is corrosive, and must be removed before the gas is transported). Natural gas or crude oil containing a high sulfur content is termed "sour", while those with low concentrations are called "sweet" (presumably because "sweet" is the opposite of "sour").

Crude oils can contain a high proportion of long-chain hydrocarbons used to make lubricating oils, waxes and asphalt. These crudes are more viscous and gooey, and are called "heavy". Conversely, crude oils with a high proportion of smaller hydrocarbon molecules are called "light".

Most crude oils produced in Australia from the Bass Strait are "light" and "sweet". Consequently, Australian refineries have traditionally needed to import lubricating oils and asphalt, or alternatively, have imported heavy crude oils needed to produce these products. It is not unusual for countries like Australia to export some of the crude oil that they produce while, at the same time, importing crude oil from overseas. This allows refineries to produce the various hydrocarbon products in the proportions required for their local market.

Although crude oil (and "non-conventional" oil sources) vary in composition, fuels and other products derived from crude oil must have consistent quality and meet specifications that apply internationally. Producing such hydrocarbon fuels is the job of an oil refinery.

The central process in an oil refinery is the separation of crude oil into various hydrocarbon components, or "fractions". The heart of an oil refinery is a distillation unit, which is used to separate hydrocarbons into fractions suitable for use as gasoline, aviation gasoline, diesel fuel, jet fuel, kerosene, lubricants, asphalt and other products.

A good five-minute video showing the distillation of crude oil can be viewed at:
https://www.youtube.com/watch?v=KCs1F_44dy4

17. Oil refining
and the birth of the petrochemical age

If you were to go back to the 1920s or 1930's and visit an oil refinery, it would have been simpler than the oil refineries we have today. Most oil refineries were located at ports, as they are today, to allow crude oil to be brought in by large tanker ships. The oil was held in storage tanks, allowing a constant supply of oil to the refinery during periods between shipments. The heart of the oil refinery was (and still is) the fractional distillation column, in which various fractions of crude oil (gasoline, diesel, kerosene, asphalt, etc) were separated according to their volatility (boiling point). Then, these products are held in separate storage tanks until they are ready to be delivered.

I didn't actually see an oil refinery in the 1930's – I was not yet born, but here's a simplified diagram showing how it worked:

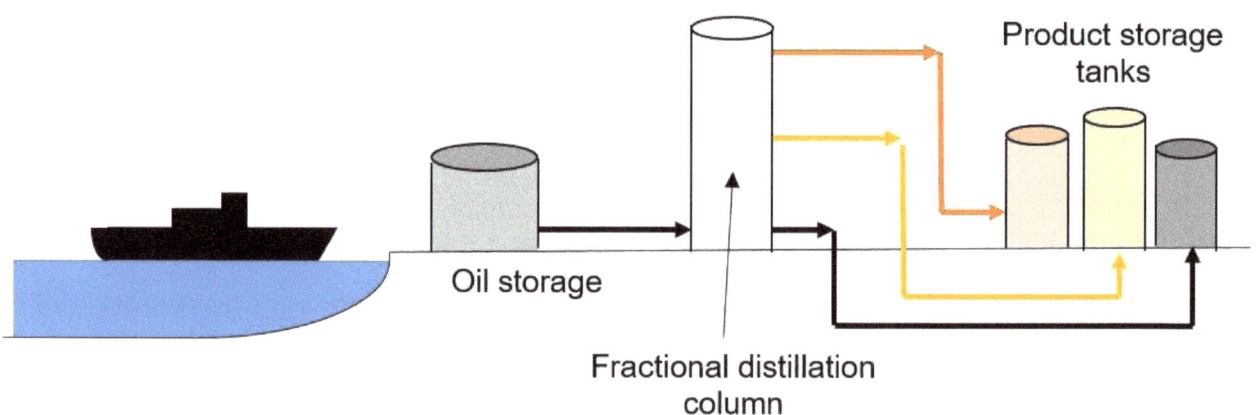

The oil industry was developing quickly to meet the fuel requirements of the rapidly-developing motor vehicle industry, and in particular, to meet growing needs for gasoline (petrol). Bear in mind that, early in the 20th century, gasoline was the primary fuel used for motor vehicles. Diesel engines did not become common until later, and jet engines were not yet invented. Cars were becoming more sophisticated than the "horseless carriages" of the early 20th century, and there was a strong drive towards more powerful engines and greater fuel efficiency.

To achieve higher efficiency and power output, it was necessary for gasoline engines to have higher compression ratios. In other words, the air-fuel mixture sucked into the engine cylinders would be compressed to a smaller fraction of its original volume before it was ignited. However, the drive for engine and car manufacturers to use high-compression engines posed a problem for oil refiners. At higher compression ratios, the air-fuel mixture would not burn evenly and smoothly as it drove the piston during the engine's power stroke. Rather, the air-fuel mixture would detonate, producing intense shock waves. This effect, called "knocking", would cause vibration and, if severe, could damage or destroy the engine.

I first started driving my father's old, poorly-maintained rustbucket cars during the late 1960s. It was not unusual for the engine to sound like it was grinding coconuts during periods of acceleration or uphill driving, indicating that you needed to ease off the accelerator pedal. With modern cars, we don't hear "knocking" because the computerised engine management system detects any such vibration, and immediately delays the ignition timing. This causes the air-fuel mixture to be ignited later, after the gas mixture begins to expand, and effectively reduces the compression ratio to a level at which the fuel burns smoothly.

The many hydrocarbon compounds in gasoline differ widely in their tendency to "knock". As it turned out, "straight-chain" hydrocarbons (those with the carbon atoms in one continuous row) tended to detonate at low compression ratios, while "branched chain" hydrocarbons allowed much higher compression ratios to be used.

For example, "normal" (or "n") heptane, whose molecular structure consists of seven carbon atoms in a row, is an extremely poor fuel for gasoline engines.

```
n - heptane    H   H   H   H   H   H   H
               |   |   |   |   |   |   |
           H — C — C — C — C — C — C — C — H
               |   |   |   |   |   |   |
               H   H   H   H   H   H   H
```

On the other hand iso-octane, whose molecular structure is highly branched, is an excellent fuel for gasoline engines. It burns smoothly without "knocking" even at compression ratios of 8:1 or higher. The molecule of iso-octane has a backbone of five carbon atoms (and thus, can be considered to be derived from a pentane molecule). Two methyl groups (a carbon atom bound to three hydrogen atoms) are bonded to the second carbon atom in the chain, and a third methyl group is bonded to the fourth carbon atom. Because there are three methyl groups bound to a backbone of pentane at carbon atoms 2, 2 and 4, it has the chemical name "2,2,4 – trimethylpentane".

```
              H            H
              |            |
     H   H-C-H   H   H-C-H   H
     |     |     |     |     |
 H — C  —  C  —  C  —  C  —  C — H
     |     |     |     |     |
     H   H-C-H   H     H     H
              |
              H
```

Iso-octane
2,2,4 - trimethylpentane

Oil refiners and engine manufacturers developed a system to rate the quality of gasoline in regard to its tendency to "knock". This measure is called the "octane rating".

Note that the octane rating applies *only* to gasoline – it has no relevance whatsoever to diesel or jet fuel, which are burned under different conditions. The octane rating is based upon the performance of a fuel compared to a mixture of n-heptane and iso-octane. As we have seen, n-heptane is a lousy fuel for gasoline engines, and was assigned an octane rating of 0. Iso-octane was the best fuel that was known at the time, and was assigned an octane rating of 100.

Fuels are tested by operating a gasoline engine at various compression ratios until knocking begins to occur. The fuel is assigned an octane rating according to the equivalent mixture of iso-octane and n-heptane that gives the same performance. So, if a fuel performs like a mixture containing 80% iso-octane, its octane rating is 80.

Standard petrol that is currently sold in Australia has an octane rating of 91-92. Some European model cars have engines with higher compression ratio, and premium fuels with octane ratings of 95 and 98 are available.

One solution that created greater problems than it solved

One cheap and effective way to enhance the octane rating of gasoline is to add tetra-ethyl lead, and this was commonly done throughout the world from the 1920s to the 1970s and 1980s. However, the lead content of fuel is emitted in engine exhaust gases, and low-level lead pollution became endemic throughout the environment. Lead was present in the air in congested cities, accumulated in soils and lakes, and found its way into the food chain. Lead is a heavy metal neurotoxin. It adversely impacts on intelligence and behaviour (learning difficulties, aggression, hyperactivity and impulsiveness), as well as causing hypertension and damage to various organs. It is a particular concern in regard to the mental development of children.

Because metallic lead is soft and pliable and readily shaped, it was used by the ancient Romans for plumbing systems that provided fresh water directly to the homes of the ruling class. (the chemical symbol for lead is Pb, from the latin word *plumbum*). Roman leaders would have ingested lead salts that dissolved in the water from corroded piping and aqueducts, and very likely suffered adverse effects of low-level lead poisoning on their mental and physical health. Of far greater impact would have been the adverse effects suffered by ordinary citizens and legions of the Roman Empire, who bore the consequences of bad decisions by rulers who were mentally impaired, or possibly even insane. It is widely believed that lead poisoning was a contributing factor to the decline and ultimate collapse of the Roman Empire.

The health impacts of leaded gasoline began to be realised as early as the 1920s, when some workers at tetraethyl lead factories became ill, insane and died. Research in the 1970s showed that low levels of lead exposure in young children caused permanent learning and behaviour problems. Extensive tests showed a strong correlations between lead levels in children's blood and brain damage, hypertension and learning disorders. The use of lead in petrol is now considered to have been one of the greatest public health disasters of the 20th century.

By the 1970s, governments realised that steps were needed to deal with the growing problems of lead pollution, as well as other air pollutants that were causing smog in cities. Lead is not only poisonous to people, but it also poisons catalysts used in catalytic converters to remove carbon monoxide, unburned hydrocarbons and nitrogen oxides from car exhaust.

Despite opposition from manufacturers of tetraethyl lead, governments in the USA, Australia and many other western countries required that new cars must be built to operate on unleaded gasoline (with hardened valve seats), and then instituted a phase-out of tetraethyl lead. Since leaded gasoline was phased out (and currently, virtually eliminated), lead concentrations in the air of major cities has dropped sharply and blood lead levels have declined markedly. Generally, air quality in many western cities has improved over the past 30 years with better emissions control technology in motor vehicles (but air quality has gotten worse in many Asian cities due to industrialisation and vastly increased numbers of cars).

Researchers from the National Bureau of Economic Research concluded that the reduction in childhood lead exposure in the late 1970s and early 1980s was largely responsible for the decline in violent crime which has occurred across the United States since the early 1990s [See Reference 1]. The rate of murder dropped by half between 1992 and 2010, although the US murder rate is still much higher than many other countries. The researchers estimate that half the reduction in violent crime could be attributed to reduced exposure to lead as children were growing up (with about 30% of the reduced crime attributed to legalisation of abortion, which avoided many children being born to parents who were unable or unwilling to care for them). If this estimate is accurate, the phasing out of leaded gasoline avoided 100,000 murders being committed over the past 20 years – more than the total number of US soldiers killed in the Vietnam, Iraq and Afghanistan wars combined!.

The phase-out and elimination of leaded gasoline in most countries of the world is one of the few major environmental success stories of our generation.

Current technology to improve octane rating of gasoline

Another way of increasing the octane rating of gasoline is to convert "straight chain" hydrocarbon molecules into "branched chain" hydrocarbons. This is accomplished by passing the hydrocarbons over specific catalysts at the appropriate temperature and reaction conditions. This process, which is called "isomerisation", has been in use since the 1940s and is the main method currently used to increase the octane rating of gasoline.

During isomerisation, some carbon-carbon bonds within the hydrocarbon chains are broken, and the molecules re-arrange to form a branched structure. For example, the straight chain hydrocarbon n-hexane re-arranges within an isomerisation reactor to form several branched-chain hydrocarbon isomers. There are five isomers of hexane (C_6H_{14}), whose molecular structures are shown below:

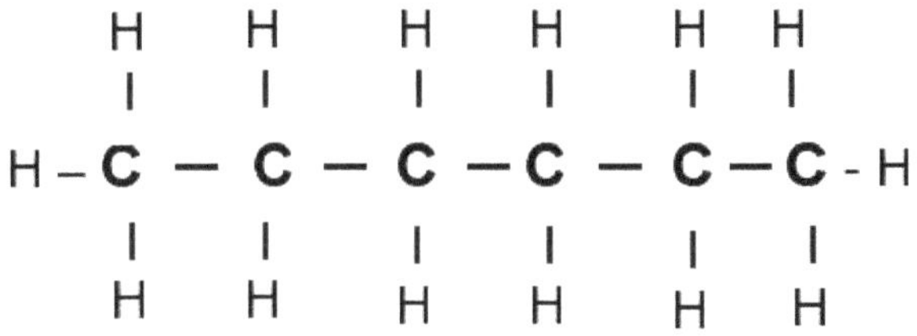

```
   H   H   H   H   H  H
   |   |   |   |   |  |
H– C – C – C – C – C– C - H      n – hexane
   |   |   |   |   |  |          Octane Number 25
   H   H   H   H   H  H
```

2-methylpentane
Octane Number = 73

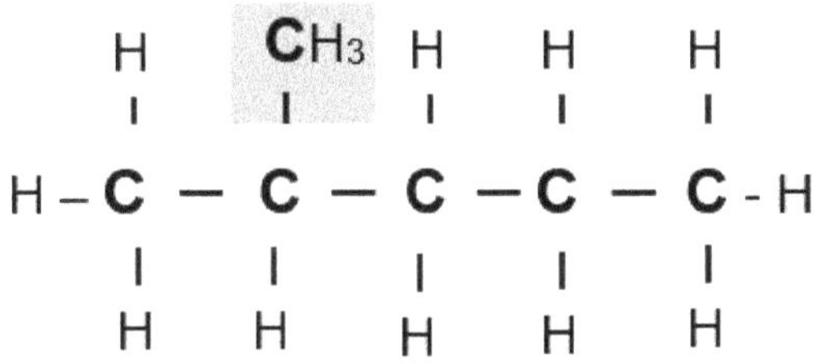

```
   H   CH3  H   H   H
   |   |    |   |   |
H– C – C  – C – C – C - H
   |   |    |   |   |
   H   H    H   H   H
```

3 – methylpentane
Octane Number = 75

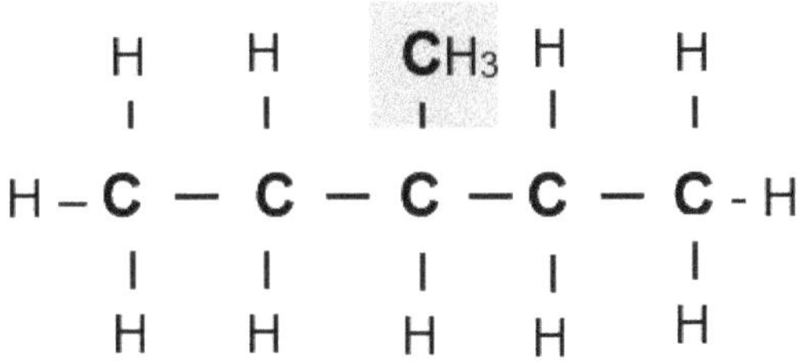

```
   H   H   CH3  H   H
   |   |   |    |   |
H– C – C – C  – C – C - H
   |   |   |    |   |
   H   H   H    H   H
```

2,2 – dimethylbutane
Octane Number = 92

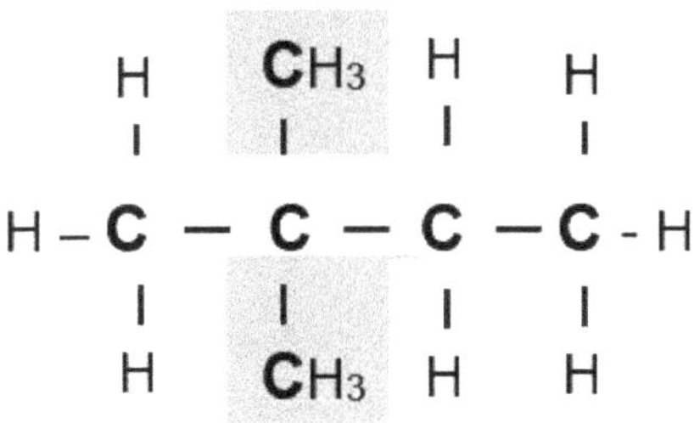

```
   H   CH3  H   H
   |   |    |   |
H– C – C  – C – C - H
   |   |    |   |
   H   CH3  H   H
```

2,3 – dimethylbutane
Octane Number = 101

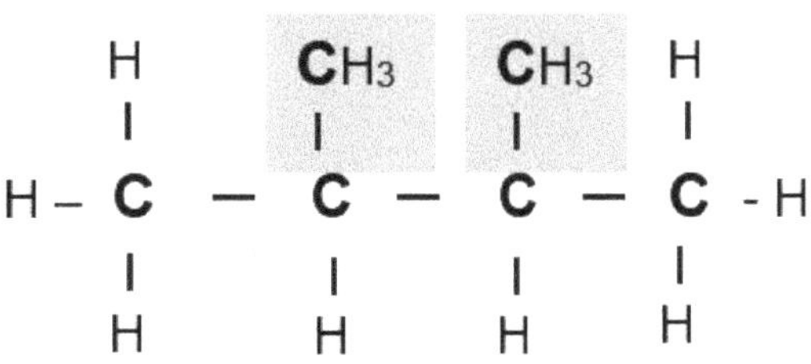

```
   H   CH3  CH3  H
   |   |    |    |
H– C – C  – C  – C - H
   |   |    |    |
   H   H    H    H
```

I shall leave as an exercise for the motivated reader to work out the structures and names for the isomers of octane (C_8H_{18}). Remember to start numbering the longest chain at the end which gives the lowest numbers for "side groups". If you find that two isomers have the same name, they are the same compound. I counted 15 isomers for octane (which are listed at the end of this chapter). You should find this exercise to be simpler, and presumably more productive, than doing a crossword or sudoku puzzle.

How we won the war

By the start of the Second World War, large amounts of gasoline were needed for military operations. Unlike Germany (which had adapted diesel engines for their tanks), the US and Allied armies used gasoline for all their aircraft, tanks and trucks. Note that military vehicles are designed for maximum performance in battle – not for fuel economy. The M4 Sherman Tank, the primary battle tank used by the United States and its western Allies, consumed 3.5 litres of gasoline for each kilometre travelled (using as much fuel as 35 cars).

The high demand for gasoline for the Allied war effort represented a major challenge for oil refineries (the Axis powers of Japan and Germany had even bigger problems getting sufficient oil supplies to maintain their war effort).

The United States, Australia and other wartime Allies needed to produce the maximum amount of gasoline, but they did not need heavier oil fractions (those that would now be used for diesel, jet fuel, kerosene, asphalt and waxes). The solution, which is reputed to have been a major factor in the Allies winning the war, was the adoption of catalytic cracking technology. This technology had been developed in the 1930s, but was largely ignored until a pressing need arose.

"Catalytic cracking" uses a catalyst, at the appropriate reaction temperature and conditions, to break carbon-carbon bonds within hydrocarbon chains. It "cracks" or splits large hydrocarbon molecules into smaller ones. The reaction is non-specific, and can split any of the carbon-carbon bonds within the molecule. Large hydrocarbon molecules can be split into two smaller molecules of roughly equal size, or the reaction might simply break off one or two carbon atoms at the end of the chain.

While refineries wanted to maximise their production of gasoline (and now, diesel fuel, as well), catalytic cracking produces substantial amounts of smaller molecules, containing one, two or three carbon atoms.

Let's consider, for example, what would happen to n-decane when a two carbon segment is cleaved by catalytic cracking.

When the carbon chain is split, there are not enough hydrogen atoms to form two alkane molecules.

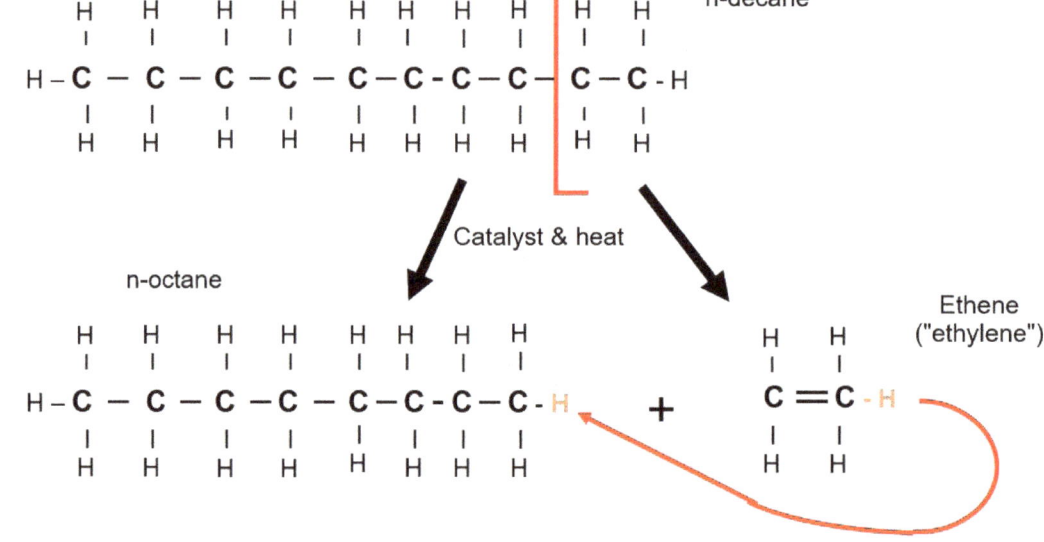

One molecule ends up with a carbon-carbon double bond, which is very chemically reactive. Chemists refer to the carbon-carbon double bond as a "functional group", as it is an active site for the molecule to undergo further chemical reactions.

Hydrocarbons containing one (or more) carbon-carbon double bond are called "alkenes" or "olefins". They are much more interesting to chemists than alkanes, which burn very nicely, but don't readily undergo other chemical reactions. Chemists can use the carbon-carbon double bond as a "handle" to manipulate the molecule to form all sorts of other compounds. The large-scale production of alkenes (like ethene, shown above being produced by catalytic cracking) opened up new possibilities for chemists to produce a wide range of new compounds, starting from petroleum. Alkenes are like chemical Lego pieces that can be used to build all sorts of chemical compounds (plastics, herbicides, pesticides, pharmaceuticals). They gave rise to the birth of the modern petrochemical industry.

Before we leave the leave the oil refinery and look at the "downstream" production of plastics and other petrochemicals, here's a simplified diagram of a modern oil refinery, including catalytic cracking and isomerisation technology

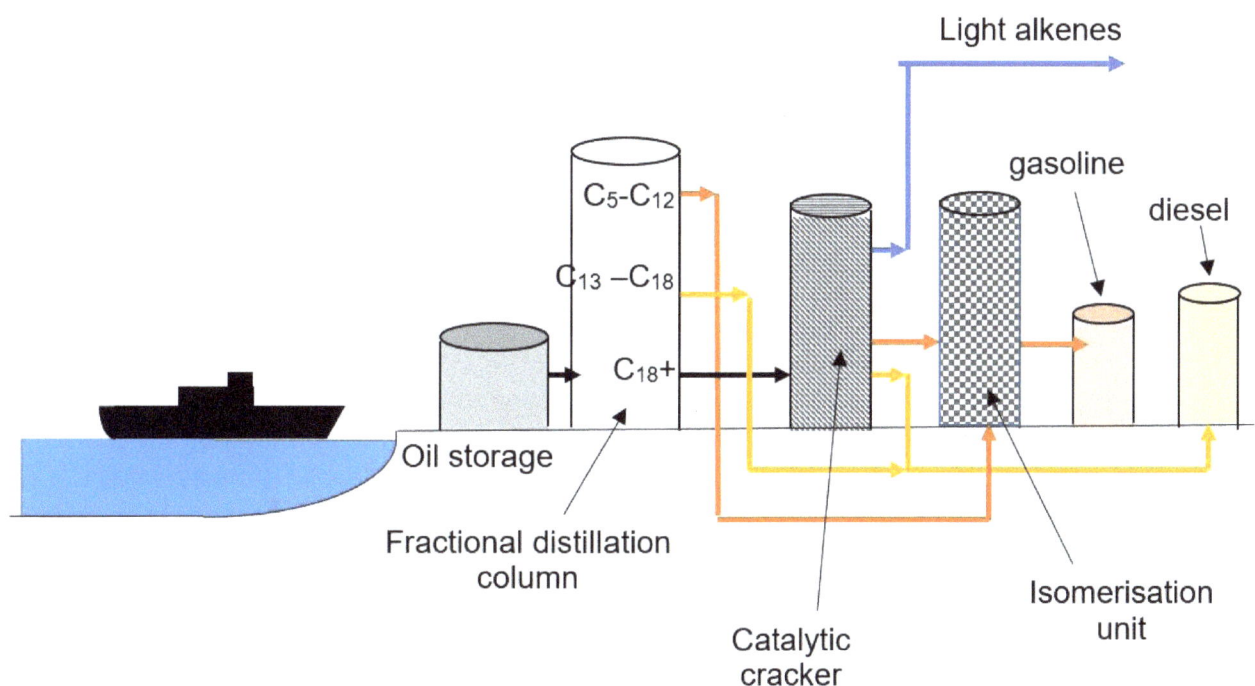

References

1. J.W. Reyes, Environmental Policy as social policy? The impact of childhood lead exposure on crime, May 2007. http://www3.amherst.edu/~jwreyes/papers/LeadCrimeNBERWP13097.pdf

Answer: Isomers of octane

n-octane, 2-methylheptane, 3-methylheptane, 4-methylheptane, 2,2-dimethylhexane, 3,3-dimethylhexane, 2,3-dimethylhexane, 2,4-dimethylhexane, 2,5-dimethylhexane, 2,2,3-trimethylpentane, 2,2,4-trimethylpentane, 2,3,3-trimethylpentane, 2,2,3,3-tetrmethylbutane, 3-ethylhexane, 3-ethyl – 2-methylpentane.

18. The age of plastics

Humans have been using stone, wood and other naturally occurring materials from the environment as long as we have been a species. In the last few thousand years, people began to fundamentally transform the materials in their environment into entirely new chemical forms. This probably began with the production of metals on a limited scale for tools and weapons, and with the fermentation of foods as a means of preservation and as an intoxicant.

Fermentation of sugar has been used to make alcoholic beverages for thousands of years. This process is undertaken by single-cell yeast organisms whose spores are naturally present in the air. All that was necessary was for people to provide a sugary solution as a food source for the yeast, and an airtight container maintained at the right temperature and conditions. The chemical reaction undertaken by the yeast cells is:

$$C_6H_{12}O_6 \xrightarrow{\text{yeast}} 2\ C_2H_5OH + 2\ CO_2$$

sugar → ethanol + carbon dioxide

The key actor in this process is the yeast, which undertakes the chemical transformation of sugars into ethyl alcohol (ethanol). You might be wondering, why is yeast so generous to produce ethanol for us. What's in it for the yeast?

The production of ethanol would have been an evolutionary adaptation that assisted yeast to survive in the natural world, where yeast compete against bacteria for food. The problem facing yeast cells is that they are relatively slow growing, which makes it difficult for them to utilise the sugar food source before bacteria do. To address this disadvantage, yeast developed the ability to produce ethanol as a chemical warfare strategy. Ethanol is toxic to bacteria, inhibiting their growth until the yeast can utilise the food source. However, once the concentration of alcohol in the solution exceeds about 12%, ethanol becomes toxic even to the yeast, and the fermentation process stops. Production of more concentrated solutions of ethanol (such as gin, vodka or whiskey) needed to await the invention of distillation, which separates ethanol from the sugary solution in which it is produced. Currently, "fortified wines" (such as sherry and port) are produced by adding extra ethanol to fermented fruit juice.

By the second half of the 20th century, with the widespread refining of petroleum, many new chemical compounds were being produced by entirely man-made synthetic processes. Many of these processes are based on small hydrocarbon molecules containing carbon-carbon double bonds. The simplest such alkene (or "olefin") is ethene (commonly called "ethylene"), containing two carbon atoms.

The carbon-carbon double bond "locks" the ethylene molecule into a flat, two-dimensional shape, as drawn on this page. The bond angles are about 120 degrees, so the axis of the chemical bonds point towards the corners of an equilateral triangle.

Ethylene is highly chemically reactive because of the presence of the double bond, and readily undergoes "addition reactions" with many compounds under suitable reaction conditions. One simple example is the industrial production of ethanol, which provides an alternative route to fermentation (which still produces most of the ethanol used today). Ethylene is reacted with steam in a chemical reactor with a suitable catalyst and appropriate conditions of temperature and pressure. During the reaction, an oxygen-hydrogen bond within the water molecule is broken, and the resulting fragments add to the carbon-carbon double bond in ethylene.

$$\begin{array}{c} H \\ \\ H \end{array} C = C \begin{array}{c} H \\ \\ H \end{array} + H_2O \rightarrow \begin{array}{c} H H \\ | | \\ H - C - C - OH \\ | | \\ H H \end{array} \text{ Ethanol (ethyl alcohol)}$$

When people refer to "alcohol", they are normally referring to ethanol, but ethanol is just one of many compounds that are called "alcohols". All contain an OH group bonded to a chain of carbon atoms.

As it turns out, ethanol is the least toxic of all the "monohydric alcohols", that is, alcohols containing a single OH group within their molecule. The simplest alcohol is methanol (CH_3OH), whose molecules contain a single carbon atom. Propanol (C_3H_7OH), which has three-carbon atoms, is widely used as antiseptic "rubbing alcohol" to kill bacteria on open wounds or to sterilise the skin before an injection. There are two isomers of propyl alcohol. 1-propanol has the OH group at the end of the carbon chain, and is therefore a "primary alcohol", while 2-propanol has the OH group in the middle of the three-carbon chain, and is termed a "secondary alcohol".

Since the OH groups imparts toxicity to alcohols, you would expect that a compound having **more than one OH group** would be **more** toxic. But, as it turns out, you would be wrong! Dihydric compounds, having two OH groups in their molecule, are less toxic. And, compounds with many OH groups tend to have no toxicity at all. Among these are the sugars and starches that are an integral part of our diet.

1 - propanol

2 - propanol

Two important alcohols having multiple OH groups are ethylene glycol (1,2 – ethanediol) and glycerol (also called "glycerine", or 1,2,3 - propanetriol). As we will see, these compounds are commonly used for producing more complex molecules.

Ethylene glycol (1,2 – ethanediol)

Glycerine (1,2,3 –propanetriol)

Alcohols are useful starting points for many chemical reactions. In particular, primary alcohols are readily oxidised to form carboxylic acids. The –CH$_2$OH group at the end of the alcohol molecule is converted to a –COOH carboxylic acid group. For example, ethanol reacts with the air to produce ethanoic acid (commonly called "acetic acid"). This explains why a bottle of wine will often go "off" within a few days after its cork is removed (allowing air to enter the bottle). The ethanol content of the wine is oxidised by the air, forming vinegar (a dilute solution of acetic acid).

$$\text{ethanol} + O_2 \rightarrow \text{Ethanoic acid (acetic acid)} + H_2O$$

Once we have produced a carboxylic acid, such as ethanoic acid, it can react with an alcohol to produce a type of chemical compound called an "ester". The esterification reaction links the –CH$_2$OH group at the end of the alcohol molecule with the –COOH group at the end of a carboxylic acid molecule. A water molecule is removed in the process. For example, ethanol will react with acetic acid to produce the ester "ethyl acetate".

$$\text{ethanol} + \text{acetic acid} \rightarrow \text{ethyl acetate ester} + H_2O$$

Ethyl acetate is a colourless liquid which, like many esters, has a strong characteristic odour. Traces occur naturally in grapes and this is responsible for the "fruity smell" of some wines. Ethyl acetate is produced in large quantities (about 1.3 million tonnes per year) by reacting ethanol with acetic acid, and is widely used as a solvent in glues, nail polish remover and to decaffeinate coffee and tea. Many other esters occur naturally in trace concentrations in various fruits, and are largely responsible for their particular smell.

In fact, substantial amounts of esters by produced by virtually all plants and animals. Many plant seeds contain a high content of vegetable oil, and these oils are often extracted as an ingredient in food and for cooking. Common examples include olive oil, sunflower oil, safflower oil and peanut oil. All of these oils are tri-esters formed from glycerine and long-chain carboxylic acids (typically containing 16 or 18 carbon atoms). So too are all animal fats. Vegetable oils differ from animal fats mainly in that their long-chain carboxylic acids contain one or more carbon-carbon double bonds. In other words, vegetable oils tend to be "unsaturated" or "polyunsaturated" (containing several carbon-carbon double bonds), while animal fats contain long-chain carboxylic acids that are "saturated".

```
                                    Long-chain carboxylic acid (oleic acid) segment
        H           O
        |           ||
    H - C - O  -  C - (CH₂)₇ - C = C - (CH₂)₇ - CH₃
        |           O
        |           ||
    H - C - O  -  C - (CH₂)₇ - C = C - (CH₂)₇ - CH₃
        |           O
        |           ||
    H - C - O  -  C - (CH₂)₇ - C = C - (CH₂)₇ - CH₃
        |
        H
                   Glycerine segment
```

A typical tri-ester molecule found in olive oil. It is formed by esterification of a glycerine molecule with three molecules of long-chain carboxylic acid (in this case, oleic acid, which is mono-unsaturated).

As well as being used as food, vegetable oils and fats can be used as fuel. I recall that one day, during the "oil crisis" in the 1970s, I viewed a bottle of vegetable oil purchased from our local supermarket, and thought that it looked very much like furnace oil. I wondered if vegetable oil could be used as fuel. I discovered later that vegetable oils and fats could indeed be used as fuel. In fact, the inventor of the diesel engine, Rudolph Diesel, had intended that his engine would operate on peanut oil (at the time, in the early 1900s, diesel fuel was not available because diesel engines had not yet been invented!).

Biodiesel and biofuels: the great dilemma

The high combustion energy of vegetable oils, and their excellent lubrication characteristics, are ideal for combustion in diesel engines. However, vegetable oils are more viscous than standard petroleum-based diesel fuels, and would not meet existing diesel fuel specifications without treatment. "Biodiesel" fuel (diesel fuel produced from biological sources) is made by chemically modifying vegetable oils to reduce their viscosity. The process splits the ester linkages within the tri-ester molecule, releasing three molecules of long-chain carboxylic acid and a molecule of glycerine. Each carboxylic acid molecule is then esterified with methanol (whose molecules contain a single OH group). This process, called "transesterification", converts each tri-ester molecule into three smaller mono-ester molecules, providing the desired viscosity. The process produces glycerine as a byproduct (which is sold for various applications), and requires an input of methanol. Animal fats, which have even higher viscosity than vegetable oils, are transesterified in the same way to produce biodiesel fuel.

Biodiesel fuel is currently used in many countries, particularly in Europe, and some countries have enacted laws requiring that diesel fuel contain a minimum concentration of biodiesel. On the surface, this seems to be an environmentally sound policy. It replaces diesel fuel made from crude oil, a limited and non-renewable resource. It greatly reduces emissions of greenhouse gases, since the carbon in the oil was initially removed from the air by photosynthesis (so that carbon dioxide is emitted and then absorbed through a closed cycle). And, indeed, producing modest quantities of biodiesel from waste animal fats and used cooking oil makes perfect sense.

However, to produce huge volumes of biodiesel fuel (required by millions of trucks, buses and commercial vehicles operating around the world) requires the allocation of vast areas of agricultural land to grow oilseed crops. By creating a huge market for biofuel, governments have diverted agricultural land and resources away from growing food and other agricultural products. This has caused increased food prices, exacerbating hunger and poverty in poor countries that rely on food imports. In some countries with poor environmental controls and endemic corruption, increased demand for vegetable oils has led to natural forests being cleared to establish palm oil plantations. Thus, high demand for biodiesel fuel has led to negative environmental, social and economic consequences.

Overall, while biodiesel can make a useful modest contribution to reducing the world's consumption of fossil fuels and reducing greenhouse gas emissions, it seems unlikely to provide most of the world's needs for transport fuel without major detrimental impacts.

The same considerations apply to other "biofuels" intended to replace petroleum-based fuels with compounds of biological origin. Another important biofuel is ethanol, which is commonly used as a replacement for gasoline in Australia and other countries. It is often used as an "E10" mixture (containing 10% ethanol and 90% gasoline), although specially-modified engines can use mixtures containing 85% (or higher) ethanol. Ethanol fuel is generally produced by fermentation of sugars extracted from various crops. It is produced mainly from sugar cane in Australia and Brazil, and from corn in the United States.

With an octane number of 108, ethanol is a valuable additive to enhance the octane rating of gasoline. Its addition also increases the volatility of the fuel mixture, which presents some challenges to refineries to meet fuel specifications for summer conditions. In principle, by displacing petroleum-based gasoline with a compound produced by photosynthesis, ethanol fuel can reduce emissions of greenhouse gases. In actual practice, greenhouse gas emissions are only partially reduced because petroleum-based fuels are consumed in distilling ethanol produced from corn, and petrochemicals are consumed in producing fertilisers and pesticides required to grow these crops.

As with biodiesel, the production of huge amounts of ethanol requires the diversion of agricultural land and resources away from food production. So while ethanol and biodiesel provide a means to reduce reliance on unsustainable use of non-renewable fossil fuel resources, it is questionable whether biofuels could ever meet most of society's needs for transport fuels without severe adverse social and environmental effects.

The synthesis of new materials

The previous examples have shown that, by utilising simple molecules as building blocks, we can produce compounds that are already available in nature, or we can modify natural compounds to better suit our requirements. However, the same process of chemical synthesis can produce entirely new materials, pharmaceuticals and other products that do not exist in nature. One particular group of such products are polymers, commonly referred to as "plastics", which are now ubiquitous. Polymers are used in nearly every appliance, manufactured component, electronic device, food packaging, furniture, paints and coatings. Modern life as we know it would be inconceivable without plastics. They are produced with an extraordinarily broad range of physical, chemical, electrical and optical properties.

One type of polymer uses the reaction of alcohols and carboxylic acids to produce esters. The "trick" is simply to use a dihydric alcohol (whose molecule contains **two** alcohol OH groups) and a dicarboxylic acid (whose molecule contains **two** carboxylic acid groups). Each molecule then becomes chemically bonded at each end to two other molecules which, in turn, are bonded to other molecules in a long chain. The end product is a molecule containing hundreds, or thousands, of molecular segments forming a very long chain. Each segment is chemically bonded at both ends by a polyester linkage. This type of plastic is called a "polyester".

The simplest example of a polyester would be the compound formed by the reaction of ethylene glycol and ethanedicarboxylic acid (oxalic acid).

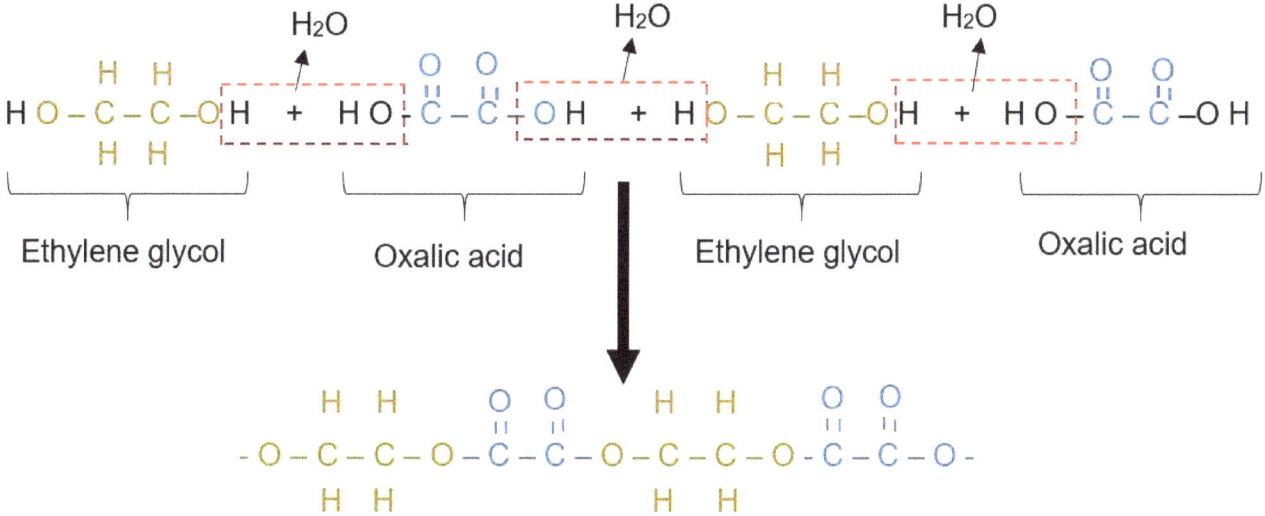

Polyesters are widely used to make fibres, films and structural components. Many of the clothing items that we wear are made of polyester-cotton blends. The polyester provides excellent durability, colour retention and crease-resistance, while the cotton provides moisture adsorption for sweat to evaporate on hot days. Boat hulls, sinks, aircraft panels, bathtubs, shower cubicles and toilet cisterns are commonly made of Fibreglass Reinforced Polyester (FRP). Glass fibres provide tensile strength for the surrounding polyester matrix. FRP was one of the first synthetic composite materials using plastics made from petrochemicals.

The long chains comprising a polyester are attracted to neighbouring chains, making the material rigid at room temperature. Heating the material allows the molecular chains to overcome the intermolecular attraction and slide past one another. The material becomes soft as it is heated, and eventually melts. Such polymers are readily formed by injecting hot, molten plastic into a die at high pressure (injection moulding), by extrusion through a die at high pressure (to form fibre, rods and tubing), or by using compressed air to blow molten polymer against the inside wall of a mould (blow moulding). These polymers are called "thermoplastics", as they are readily shaped when heated.

On the other hand, it is also possible to prepare polymers that remain rigid at high temperatures. These "thermosetting" plastics do not melt even when heated to very high temperatures; in a fire, the molecules decompose, leaving a carbon char residue.

Thermosetting polymers are produced by forming chemical bonds between adjacent polymer chains. In other words, we say that the polymer chains are "cross-linked". These "cross links" lock the molecular chains in position so they cannot slide past one another. In the previous example, if the *tri*-hydric alcohol glycerol (rather than *di*hydric ethylene glycol) were used in the polymerisation reaction, the polymer chain would include extra OH groups. The OH groups on adjacent chains could then react with molecules of ethanedicarboxylic acid to form a bridge between the polymer chains, locking them in position alongside each other. The molecular chains would form a three-dimensional rigid structure which does not soften or melt on heating.

Polyesters account for about 18% of all man-made polymer produced throughout the world, with about 50 million tonnes produced each year (coming second only to polyethylene and polypropylene in quantities produced).

The most common polyester is the thermoplastic Polyethylene Terephthalate (PET), which is made from ethylene glycol and terephthalic acid, a dicarboxylic acid. PET was invented in 1941, and is extensively used for synthetic fibres and plastic bottles (because it provides an excellent barrier to moisture and air, as well as strength and impact resistance).

The properties of PET and other polyesters can be varied by adding "copolymers" or modifying the reaction conditions. Mylar, a polyester film onto which a thin layer of aluminium is deposited, was originally developed for NASA in the 1950s, and is one of the first "space-age" materials. Mylar has been used in every manned space mission and on thousands of satellites. Polyester fibre and aluminised mylar film were among the various synthetic materials used in the spacesuits for astronauts on the Apollo moon missions.

Thermosetting polyesters are commonly used to fabricate boat hulls and other structures made of fibreglass reinforced polyester. First, a polyester resin is made by esterifying a dihydric alcohol with a dicarboxylic acid containing a carbon-carbon double bond. The resin, consisting of long polyester chains containing carbon-carbon double bonds, is a thick gooey liquid. The resin is sold to boat builders and other fabricators. They mix the resin with another compound, a copolymer, which cross-links adjacent polyester chains by bonding to the carbon-carbon double bonds.

19. Mimicking nature with man-made materials

We had previously seen that relatively small molecules containing alcohol or carboxylic acid groups could react to form larger molecules, linked by an ester group. If molecules of each starting material contain two or more alcohol or carboxylic acid groups, the reaction produces long chains of the original "monomer" units linked together. In this case, the resulting polymer is a polyester.

Alcohols and carboxylic acids are not the only kinds of molecules that react in this way. Various functional groups react to join smaller molecules together.

Another important class of plastic compounds are the polyamides, which include nylon, kevlar and many similar products. Nylon was invented in the 1930s, and became important during the Second World War as a substitute for silk and hemp in parachutes, women's hosiery, tyre reinforcing, ropes, tents, ponchos, etc.

The production of nylon is based on the reaction of molecules containing a carboxylic acid group (**-COOH**) with molecules containing an amine (**-NH$_2$**) group. A simple example is the reaction of ethanoic acid (acetic acid), which is familiar as the key ingredient in vinegar, with ethyl amine. During the reaction, these two small molecules are combined through an "amide linkage" (similar to the "ester linkages" in long-chain polyester molecules). A molecule of water is also formed by the reaction.

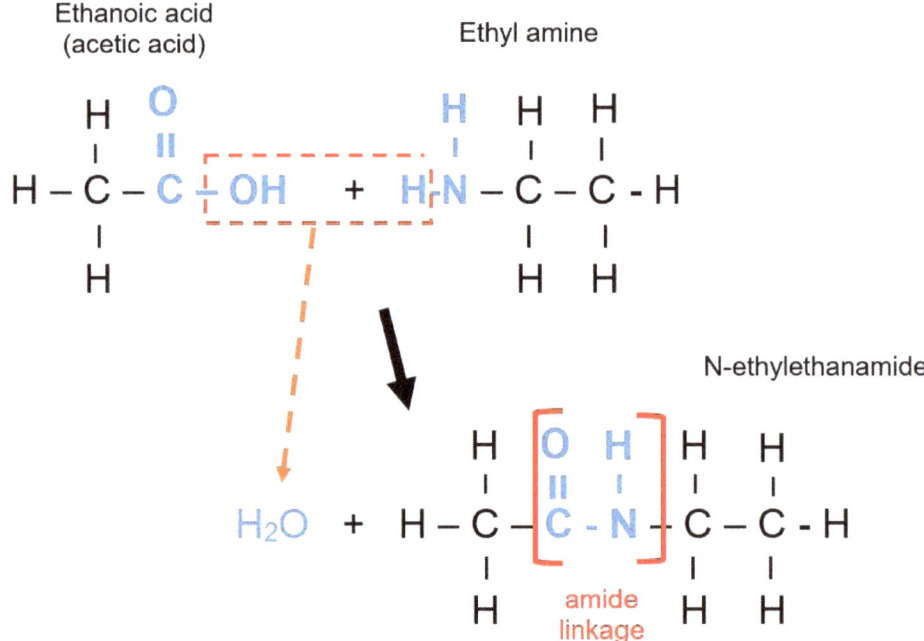

The product is a molecule comprised of two smaller "monomer" units that are chemically bonded through an "amide linkage".

If we had started with molecules containing two (or more) carboxylic acid groups, and two (or more) amine groups, the reaction would proceed to form a chain containing hundreds or thousands of these monomer units. The reaction would produce a "polyamide".

For example, we could make a polyamide by reacting oxalic acid (containing two carboxylic acid groups) with ethylene diamine (with two amine groups).

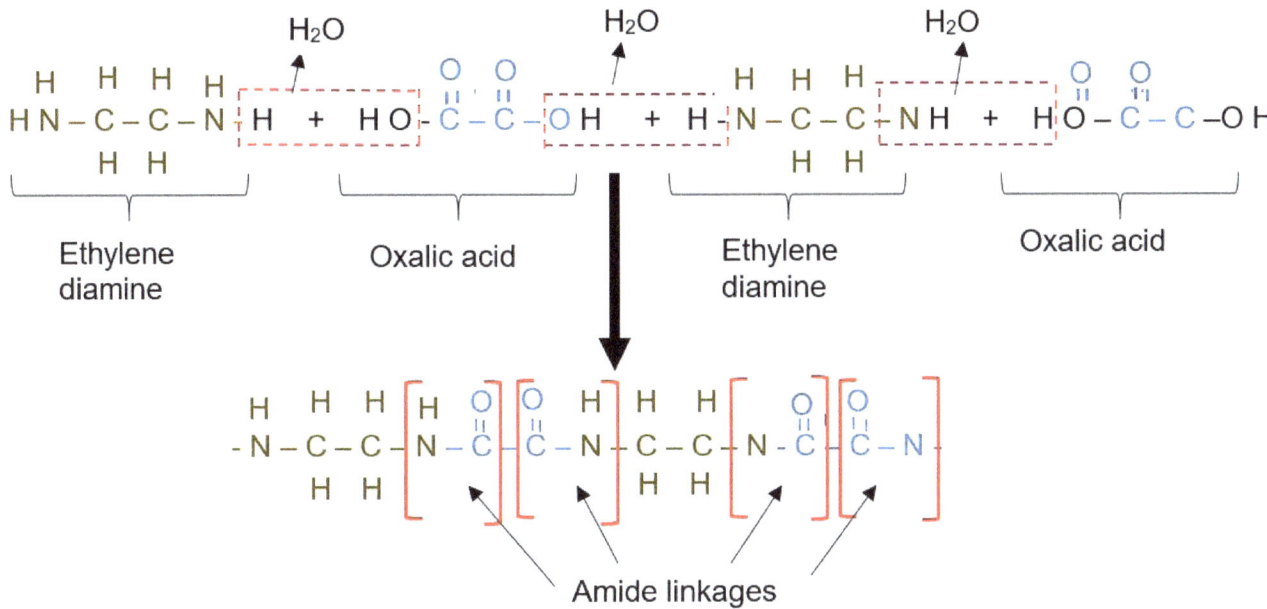

A common polyamide is made with adipic acid (which has a carboxlic acid group at each ends of a six-carbon chain) and hexane-1,6-diamine (which has amine group at each end of a six carbon chain). Because there are 6 carbon atoms in each of the reactants, this plastic is called "nylon 6, 6".

Nylon 6,6

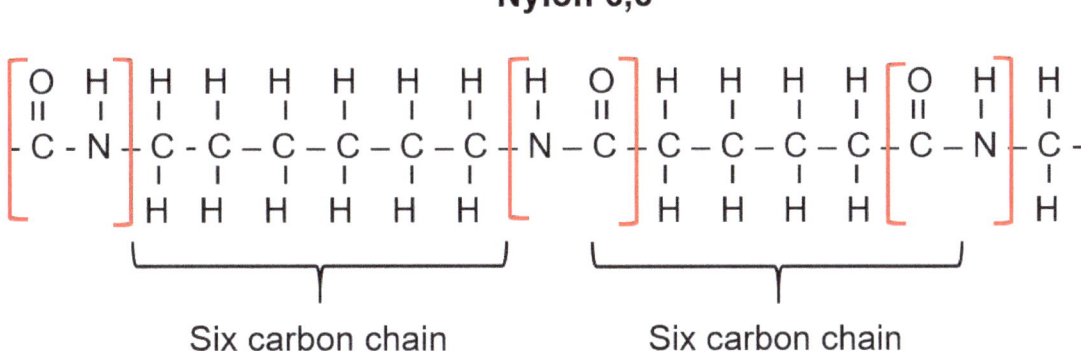

The synthesis of nylon 6,6 can be demonstrated very easily, using only very simple equipment, and is commonly demonstrated in 1st year university chemistry classes. Two short videos (about 4 minutes each) demonstrating the production of nylon fibre can be viewed on-line:

https://www.youtube.com/watch?v=yFEHKRdXb9Y

https://www.youtube.com/watch?v=NQpTQFGKRN8

Polyamides, like nylon, are the synthetic (man-made) equivalent of the protein molecules that are a major component of our bodies, as well as all animals, and even plants. Proteins are the main component of skin, muscle, tendons, nerves, enzymes, antibodies and hormones. Proteins contain long chains of molecular segments joined together by amide linkages, but there are fundamental differences between man-made polyamide polymers and proteins.

Proteins – nature's version of polyamides

Consider the structure of nylon 6,6, shown above, which is a typical synthetic polymer. It is made up of *identical repeating units* joined by amide linkages. Aside from the amide linkages, there are *no functional groups* within the chain.

The opposite situation applies for proteins found in nature. Plant and animal proteins are made of long chains, but there is enormous variety in the segments that are joined by amide linkages. There are about 20 different amino acid segments that comprise the proteins in all plants and animals, and these segments occur in a very specific order along a protein chain. The sequence of these amino acid segments is determined by the DNA comprising the genes of the plant or animal species. Our genes set out an exact blueprint for constructing every protein molecule in our body. The order of the segments along the protein chain is strictly followed when proteins are made within each cell of an animal or plant.

As a consequence, even for a small protein composed of, say, 100 amino acid segments, there are *trillions* of possible protein molecules that can be made – each having a different structure and properties. For example, there are 20 possible options for the first amino acid segment, 20 possible options for the second amino acid segment, 20 possible options for the third. That's already 20 X 20 X 20 = 8,000 possible combinations after only three amino acid segments, 64 million options after 6 amino acid segments, and about 500 billion ways to assemble nine amino acid segments.

The 20-or-so amino acid segments vary widely. Some are relatively small and simple molecular segments (the amino acid glycine contains only two carbon atoms), while others contain as many as 15 carbon atoms and have a number of functional groups. Some amino acid molecules contain more than one carboxylic acid group; some contain more than one amine group. Most contain a single chain of carbon atoms; others have rings containing five or six carbon atoms. As well as carboxylic acid and amine functional groups, amino acids may contain alcohol OH groups, ether groups, sulfur, bromine or iodine atoms. These functional groups can bind and attract other groups on the same chain, causing the protein molecule to fold or bend into a very specific shape.

The shape of protein molecules is often intricate and complex, and the three-dimensional shape of a protein is often integral to its function. A protein molecule can have the exact shape required to bind to a specific molecule, in the same way that a key fits into a lock, to cause (or to block) a specific chemical reaction. In this way, proteins often act as very specific catalysts, causing one particular chemical reaction to occur.

The amino acids from which proteins are made are different from the chemicals used to produce synthetic polyamides. To produce nylon, for example, chemists react two chemical compounds – one reagent contains two carboxylic acid groups, and the other reagent contains two amine groups. However, the amino acids from which proteins are made consist of molecules containing *both* a carboxylic acid group *and* an amine group *in the same molecule*. Consider, for example, the structure of glycine, the simplest amino acid.

$$H_2N-CH_2-COOH \quad \text{amino acid glycine}$$

Because amino acids contain **both** a carboxylic acid group and an amino group **within the same molecule**, any amino acid can form an amide linkage with any other type of amino acid. This allows for the enormous (virtually unlimited) diversity of protein molecules that can be produced from the 20 naturally-occurring amino acid building blocks.

For the amino acid glycine, whose structure is shown here, the carboxylic acid group (shown in blue) and the amino group (shown in brown) are located on adjacent carbon atoms within the amino acid molecule. In the case of glycine, there are only two carbon atoms, but it turns out that **all** 20 different types of amino acids contain a carboxylic acid and amine group on **adjacent carbon atoms** at the end of the molecule. Thus, all amino acids are said to be "alpha amino carboxylic acids". In retrospect, this should not be surprising (although the reason didn't occur to me for about forty years). If the carboxylic acid and amine groups were spaced further apart along the molecule, the carbon chain could fold and bring the two groups together, forming a closed ring structure with an internal amide linkage. The amino acid could react with itself, rather than with other amino acids. This cannot occur with the amine and carboxylic acid groups on adjacent carbon atoms.

Since our bodies are made largely of protein, we need a constant source of the 20-or-so amino acids to grow, and to repair tissues that are injured or need replacement (we are constantly growing new blood cells, skin cells, enzymes and antibodies). We, and many animal species, have lost the ability to produce a number of essential amino acids, and we must get these amino acids from the food we eat. Foods like meat, fish, eggs, yeast and moulds, and plant seeds and nuts contain a high content of protein, and these plant and animal proteins are comprised of the same 20 amino acids that make up our bodies.

All living species on Earth are assembled from exactly the same amino acid building blocks. We humans are totally dependent on the plant, animal and microbial species that we eat to provide the amino acids that our bodies cannot produce. When we digest food, the proteins are disassembled into the constituent amino acids in our stomach and intestines. The amino acids are distributed through our bloodstream to various cells in our body, where they are re-assembled in the strict sequence that is dictated by the DNA comprising the genetic blueprint encoded in our chromosomes.

Some excellent short videos discussing protein structure and function are given below. You may find some repetition among the four videos, but that might be helpful in explaining the concepts:

https://www.youtube.com/watch?v=qBRFIMcxZNM	(I = Capital letter "I")
https://www.youtube.com/watch?v=KH-LQSr7rHs	(- hyphen)
https://www.youtube.com/watch?v=MG8ziGyattk	
https://www.youtube.com/watch?v=2Jgb_DpaQhM	(_ underscore)

Of course, food provides our bodies with more than protein. We also need energy to move our muscles, to maintain essential body functions (respiration, blood flow, temperature regulation), and to think (in relation to its size and mass, the brain uses an extremely high proportion of the energy we consume). An average adult consumes about 10 million joules (2,500 kilocalories) of food energy each day. While we can derive energy from protein, and this may occur in people who are underfed, this is an inefficient use of a premium resource. A well-balanced diet should provide sufficient carbohydrate, oils and fats to provide the energy requirements for our bodies, so that protein can be reserved for those functions that it is uniquely able to do – provide for the growth, repair and regeneration of tissue.

The challenge of feeding the world's population

In the 1960s and 1970s, it was widely recognised that many people in Africa and other poor regions suffered from malnutrition. The problem was thought to be a lack of protein. A number of companies and institutions set out to develop processes to convert carbohydrate (from various grain crops) into a protein-rich product, with a texture and taste similar to meat. The aim was to identify and culture a microorganism that could readily metabolise starches and sugars, and would rapidly grow into filamentary structure (similar to meat) containing high protein content. A fungus called Fusarium venenatum was found to meet all requirements and was put into limited commercial production. In 1980, I visited the factory in the U.K. where it was being produced. I viewed the large fermentation tanks where the microorganisms were grown, and tasted samples of the product. I was very impressed. With the addition of flavouring agents, small chunks of the textured filamentary protein had a look, smell and taste that was virtually indistinguishable from chicken meat. The product was originally called "single cell protein", but later it was realised that many moulds do not live as separate, independent cells, but form into fibrous structures. Consequently, this high-protein meat substitute was renamed "mycoprotein".

Mycoprotein is commercially available, but is difficult to find (in Australia, it is sold under the trade name Quorn). It never really caught on. Why? For one thing, it turned out that malnourished people in Africa were not suffering from a lack of protein (not directly, anyway), but from a lack of food! Because their diet didn't provide enough carbohydrates, fats and oils to meet their body's immediate energy requirements, their bodies were metabolising the protein that they ate to provide energy. Their protein deficiency was not directly caused by a lack of protein in their diet, but by a lack of food calories in their diet. So, conversion of carbohydrate into protein (such as mycoprotein) was not an effective solution to the problem of malnourishment caused by insufficient food calories in the diet.

Perhaps mycoprotein also didn't catch on in western countries because of the "yuk" factor associated with eating a fungus (mould). As I'll explain, this is quite irrational, but it seems that humans are more far strongly swayed by cultural factors, aesthetics and habit than by rationality in their food preferences. Most westerners and Asians eat mushrooms as a regular ingredient in our diet, and mushrooms are fungal species that convert dead plant matter (largely carbohydrate) into protein-rich microbial matter. Most of us also enjoy yoghurt, or drink beer, in which the protein content of the food is augmented by yeast or other microorganisms. Fermentation is widely used by Asian societies to augment the nutritional content, texture and flavour of soybean curd, fish oil, and other food products. In many countries, various insect species are eaten as a regular part of the diet, and may even be considered a delicacy. Insect species are nearly as efficient as moulds at converting plant matter into protein, and grow very rapidly. A number of researchers, institutions and entrepreneurs are looking to "farm" ants, beetles, locusts and other insect species to produce high-protein flour and other protein-rich food products.

A key issue for food security in the future is how the world will increase production of meat and high-protein foods to meet the rising expectations of the emerging middle class and increasing populations in developing countries. These populations will want to improve the nutritional value and variety in their diet, at the same time as production of many food sources is being challenged. Many fish stocks in the ocean are already over-exploited, and may not be able to sustain existing production of seafood, let alone provide for increased production. There is little scope to increase the total arable land area for food cultivation, and growing competition for land to be used for biofuel production and suburban development on the outer fringes of population centres. Existing crop production is threatened by problems of soil salinity, changing rainfall patterns due to climate change, insect pests developing resistance to insecticides, and weed species developing resistance to herbicides.

On the other hand, there is potential for new plant varieties to be developed (using either genetic modification or conventional plant breeding methods) that have higher yields per hectare, higher levels of essential amino acids or greater resistance to insect pests or fungal diseases.

Some commentators argue that society must move away from meat production and consumption, since growing animals requires a far greater input of land, water and other resources than does growing food crops. In some cases, animals are grazed on marginal land which is only suitable for growing grass (with insufficient rainfall for cultivating crops). However, in many cases, livestock are raised in intensive feedlots where they are fed corn and grains that would otherwise be available for human consumption. Cattle are very inefficient at converting vegetation into meat, requiring between 5 and 20 kilograms of dried plant matter to produce each kilogram of meat. Growing pigs and poultry gives better "food conversion ratios", but even so, a high proportion of the caloric energy content of grain fed to animals is lost in the process.

Some researchers are working to develop the capability to grow animal muscle tissue (meat) artificially, converting carbohydrates into meat in a factory, without the need for live animals. To me, this has an aura of "Brave new world" or "Frankenstein". On the other hand, it could be argued that the existing practice of raising livestock animals in intensive feedlots and sending them to abattoirs is no more humane. This is an area in which ethics, culture, technology and religion intertwine. Science and technology can provide options, but it is up to us – as consumers and through the governments we elect – to decide which options are acceptable or desirable solutions to the challenges we face in the 21st century.

20. The Earth as a giant heat engine

In 1969, I had the extraordinary opportunity (along with perhaps two billion other people around the world) to watch Neil Armstrong and Buzz Aldrin take the first exploratory steps on the surface of the moon. Countless generations of our ancestors peered into the night sky and could only imagine what the moon would be like. Then, for the first time in history, we could actually see close-up photographs and video taken on the moon's surface.

As exciting as this was, the over-riding impression to me (and no doubt, to many others) was that the moon was a lifeless, dead place. Of course, I did not expect to see trees, grass, bubbling creeks and birds chirping. I knew that the moon had no atmosphere and no water on the surface, but I didn't appreciate that the mechanisms that operate on Earth to constantly redistribute heat around the planet – winds, ocean currents, evaporation, rain, storms – cannot operate on the moon. As a result, the surface of the moon is a silent motionless place, despite extreme conditions of hot and cold. High temperatures and bright sunlight in sun-exposed areas co-exist alongside freezing temperatures and darkness in shaded areas.

The same bleak lifeless scenes appeared when the first photographs were sent from the surface of Mars by the Viking 1 spacecraft in July 1976. As it turned out, Mars does have a very thin atmosphere (about one-hundredth as dense as Earth's) and occasional dust storms occur.

The surface of the moon is a silent, motionless, lifeless place. Without the action of wind or rain, this surface panorama (taken by Apollo astronaut Buzz Aldrin) had probably not changed for many millions of years. Source: NASA, http://www.hq.nasa.gov/alsj/a11/AS11-40-5913.jpg

By contrast, the Earth is literally alive with motion and life. Anywhere you may travel on the surface of the Earth, you are likely to encounter grass, trees, vegetation, insects, birds, flowing water, clouds drifting across the sky, wind and storms. Even in the most inhospitable places on Earth – deserts or frozen tundra, there is almost certainly likely to be microbial life. So, is the Earth entirely unique in the universe, or is it only one of many (millions?) of planets with life?

Before we can consider whether there are any other planets with life, we have to consider whether there are any other planets outside our solar system. Is our sun entirely unique in having planets, or are there billions of stars with orbiting planets? Until the mid-1990s, scientists had no idea. Planets are so small compared to stars, and the nearest stars are so far away, that it is not possible to directly view planets orbiting other stars, even with the most powerful telescopes. However, in recent decades, astronomers have developed ingenious and very sensitive techniques to detect other planets by looking for subtle variations in the intensity of the light reaching us from distant stars over long periods of time. Since such techniques were developed, hundreds of planets have been located orbiting other stars. In fact, astronomers have found planets orbiting most stars that they have examined with these techniques. There is now little doubt that there are millions – perhaps billions – of other planets within our galaxy.

But most such planets could not sustain life as we know it. Some are "gas giants" like Saturn and Jupiter, which do not have a solid, rocky surface. Some orbit extremely close to their star, and would be far too hot for living things to survive. Others orbit so far from their star that the surface would be too cold for living organisms. Some are in highly eccentric orbits, so that temperatures swing from extremely cold to extremely hot.

Life as we know it can only exist within a limited temperature range. In the past few decades, scientists have discovered microorganisms living in the most harsh, extreme conditions on Earth – in boiling hot or highly acidic spring water, and within the pores of rocks deep underground. Because these microorganisms survive (and apparently thrive) in such extreme conditions, they are called "extremophiles". But even these organisms only survive in places where liquid water is present, at least some of the time. Since all life as we know it requires liquid water to carry out basic functions, scientists believe that life can only exist where liquid water is present. This implies that life can only exist on planets that are not too hot, and not too cold, for liquid water to exist. This means that a planet hosting life would need to be in an orbit that is not too close to its star, and not too far . . . but just the right distance, in the so-called "Goldilocks zone".

Of hundreds of planets that have been discovered orbiting other stars, a few seem to be within the Goldilocks Zone. But this doesn't necessary mean that life is present, or even that that living organisms can survive on these planets. After all, our moon is located about the same distance from the sun as is the Earth, yet it is a barren, motionless, lifeless place. As well as being within the Goldilocks Zone, other requirements need to be met if a planet (or a moon) is to host life. For one thing, just because a planet is located within the Goldilocks Zone doesn't mean that liquid water can exist on its surface. On the moon, daytime temperatures are well above the boiling temperature of water, and night-time temperatures are well below freezing. Something else is happening on Earth, that doesn't happen on the moon, that allows the Earth to sustain the conditions for life.

The Earth is a huge "heat engine". It uses temperature differences (mainly between the equator and the polar regions) to drive billions of tonnes of air and water across the surface of the planet. Anyone who has experienced a cyclone or hurricane would appreciate the immense physical power of the wind. But even on a normal day, huge amounts of mechanical work are expended in moving air, evaporating and condensing water, and driving ocean currents around the planet. The conversion of some of the sun's heat into mechanical power provides the conditions required for life as we know it on Earth. If only a tiny fraction of this kinetic energy contained in wind could be tapped, it could meet all of humankind's need for energy.

Land masses and oceans near the equator are heated by sunlight striking nearly perpendicular to the surface, while the polar regions are constantly radiating and losing heat into the darkness of space. The equatorial region of the Earth acts a "heat source", while the polar regions serve as a "heat sink". The temperature difference between equator and polar regions affect the layer of air covering the planet, which act as the "working fluid" of a heat engine. Air near the surface at the equator is heated, expands and rises. Most incident solar energy is absorbed by the oceans, causing water to evaporate and water vapour to be carried along with the rising air stream. Heated moisture-laden air rises towards the top of the troposphere (at 10-20 kilometres elevation) and heads northwards (or southwards) away from the equator towards the poles. For reasons to be discussed shortly, the air makes its journey towards the poles in several steps – each time, cooling, descending towards the surface and dumping some of its moisture before – once again – being heated, evaporating water, and rising towards the troposphere, and continuing on its journey towards the north or south pole.

The transfer of heat from the equator towards the poles is a self-regulating mechanism. Movement of air masses and ocean currents redistributes heat from the equator to the poles, reducing the temperature differences which drive these heat transfer processes. Like all heat engines, the fraction of the heat energy which can be converted into mechanical work is proportional to the temperature difference which drives it.

The average temperature difference between the equatorial and polar regions is about 30°C, which is about one-tenth the absolute temperature of the "heat source" at the equator (relative to absolute zero). Consequently, at an absolute maximum, one-tenth of the solar energy absorbed by the Earth can be converted into the mechanical work of the wind, waves, ocean currents, rains and lightning. The remainder is radiated and lost into space. But since the total amount of solar energy absorbed by the Earth is enormous, huge amounts of mechanical power can be unleashed to sculpt the surface of the Earth, erode mountains and provide a continuous supply of fresh water and conditions for life and human society to survive and flourish.

How the Earth works

The Earth is a giant heat engine, utilising the relatively high temperatures generated by intense sunlight near the equator as an energy source to produce work. Some of this heat energy is converted into the kinetic energy of moving air (wind energy). Some generates ocean currents, moving huge volumes of water through the oceans. Some evaporates water from the oceans, with the resulting water vapour condensing into rain. The continual cycle of evaporation and condensation provides all the fresh water on Earth.

Only a small fraction of the solar energy absorbed by the Earth's surface is converted to work (producing winds, ocean currents and rain). As for any heat engine, heat that is not converted into work must be disposed to a "heat sink". In this case, the heat sink is the vast cold expanse of outer space, into which heat is radiated as long-wavelength infrared radiation.

The difference in temperature between the equatorial region and the poles drives global winds, clouds, rainfall other natural processes on Earth.

Whenever two areas on the Earth's surface are at different temperatures, this causes the air (or oceans) to move in a circular path called a "convection cell". This occurs on a global scale, driven by the temperature difference between the tropics and polar regions, but also occurs on a more local level. In particular, the oceans have a huge heat capacity, and their temperature hardly varies from one day to the next. On the other hand, land masses tend to heat up quickly during the day and then cool rapidly in the evening. This means that, during summer days, coastal land masses may get ten degrees hotter than the nearby ocean. This causes air above the land to be heated, expand and become less dense. As the column of hot air rises, it draws in air horizontally from above the ocean surface. At the same time, cooler air above the ocean is drawn downwards towards the surface creating a rotating cell of circulating air. This accounts for the cool sea breezes that are commonly experienced along the seashore.

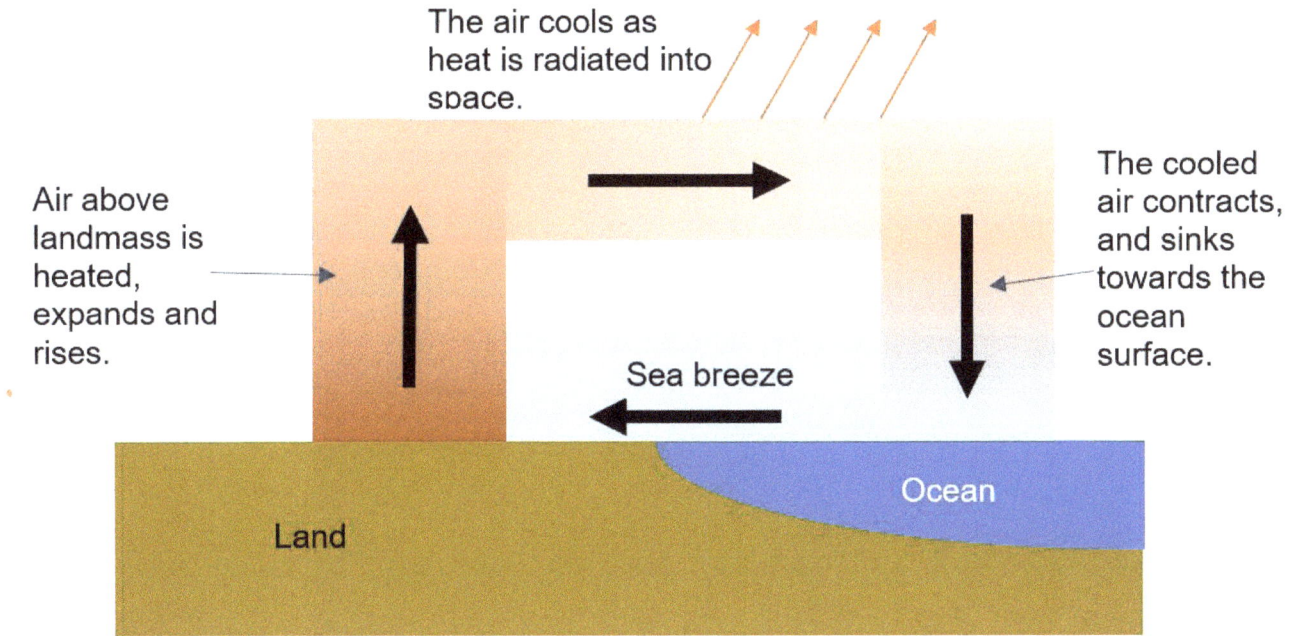

The same type of convection cell is created on a global scale, operating on the temperature difference between the tropics and the polar regions. Heated air near the equator rises through the troposphere (up to about 20 kilometres elevation), and heads towards the poles, where cold air falls towards the surface. But something else happens that interrupts the journey of these air masses across the globe. The "Coriolois Effect" comes into play, diverting the air mass once it starts its journey from the equator to the poles.

The Coriolis Effect arises from the spinning of the Earth on its axis. Since the Earth turns one revolution (about 40,000 kilometres along the equator) every 24 hours, any object at the equator is moving from west to east at about 1,600 kilometres/hour (faster than a jet airliner). However, at higher latitudes (away from the equator), the circular distance around the Earth is less, so the rotational velocity along the surface is less.

Imagine that we fire a projectile from a huge cannon at the equator, aimed directly north. The projectile will travel northwards at its launch velocity in the northerly direction. However, the projectile will also have a velocity component of 1,600 kilometres/hour towards the east. The target, located at say 30 degrees north latitude, will also be moving to the east, but its easterly velocity will be less. In fact, the target's easterly velocity will be reduced by a factor of the cosine of the latitude angle. At 30° latitude, the easterly velocity on the surface is 0.86 X 1,600 = 1,376 kilometres/hour (where 0.86 is the cosine of 30°). So, when the projectile reaches 30° latitude, it will have an easterly velocity of 1,600 – 1,376 = 224 kilometres/hour *relative to the Earth's surface*. As the projectile travels north from the equator, its trajectory veers to the right, and the projectile would land far to the east of where we aimed the cannon.

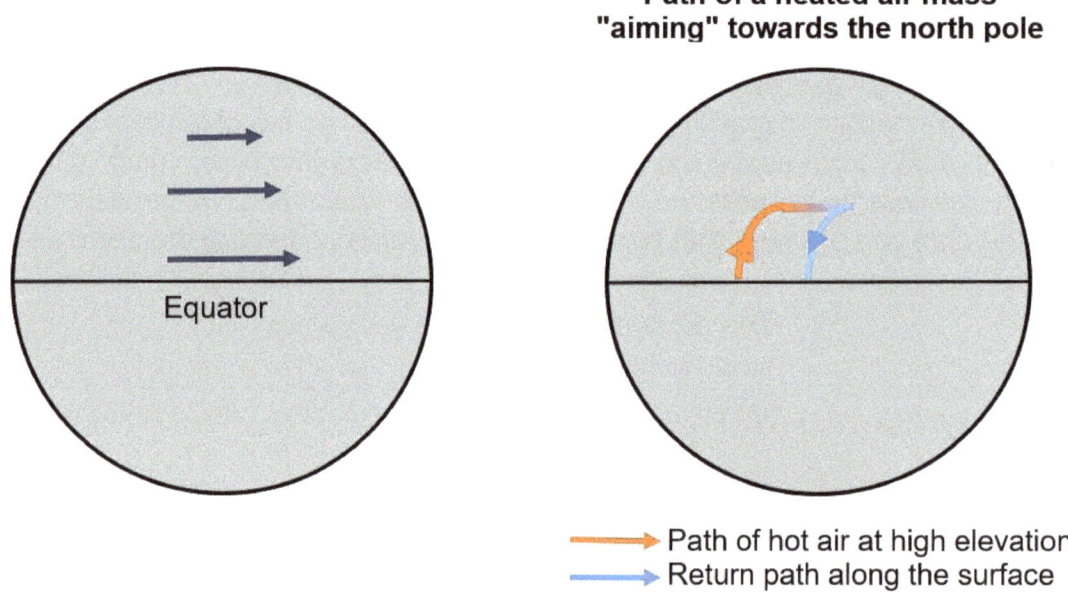

A hot air mass rising from the equator experiences exactly the same effect as it heads towards the north (or south) pole. It "aims" for the north pole, but before long, it begins to veer increasingly towards the east. By the time the air mass reaches 30 degrees latitude, it is moving virtually due east. This is about as close as this air mass can get to the north pole, so it eventually cools, sinks down through the atmosphere, and returns back towards the equator near the Earth's surface. This convection cell, from the equator to about 30 degrees latitude (occurring in both the northern and southern hemispheres) is called a "Hadley Cell". Another convection cell (called a "Ferrel cell") occurs at mid-latitude, transferring heat from about 30° to 60° latitude. And then a third convection cell occurs near the polar regions (not surprisingly, called a "Polar cell"), transferring heat on the final leg of its journey from the equator to the poles. These convection cells determine global air movements across the Earth, giving rise to the trade winds and the Jetstream.

The Coriolis Effect is a major factor in the formation of storms, and accounts for the clockwise rotation of storms ("hurricanes") in the northern hemisphere, and the counter-clockwise rotation of storms ("cyclones") in the southern hemisphere. Such storms tend to form in the late summer, where an area of the ocean has become hotter than its surroundings. Above these areas of ocean, warm humid air expands and rises (cooling as it rises, causing water vapour to condense into rain), creating a low-pressure area at the surface. "Nature abhors a vacuum" and the surrounding air starts to rush in towards the centre of the low pressure area. In the northern hemisphere, air rushing inwards from the south is deflected to the east by the Coriolis Effect, while air rushing inwards from the north is deflected to the west. As a consequence, the inwards-rushing air doesn't actually reach the centre of the low pressure area (which becomes the eye of the storm), and forms a rotating vortex around the eye of the storm.

A good description of the Coriolis Effect is provided in a 3-minute video:
https://www.youtube.com/watch?v=i2mec3vgeaI

If you wish, there is another 3-miniute video explaining the Coriolis Effect:
https://www.youtube.com/watch?v=rdGtcZSFRLk

Some other good videos are:
- Global atmospheric circulation, Covers a lot of information very quickly, 2 minutes
    https://www.youtube.com/watch?v=Ye45DGkqUkE
- Global circulation, Hadley Cells, good graphics and explanation, 5-1/2 minutes
    https://www.youtube.com/watch?v=fuw8wOhAjOg   (O = Capital letter O)
- Winds, 8 minutes
    https://www.youtube.com/watch?v=fYfrWLhZy1A

The path of the sun across the sky through the day and the seasons

One of the few things in life which is stable and predictable is the apparent path of the sun across the sky. Barring any cataclysmic events, we can work out exactly where the sun will be in the sky for any location on Earth, for next month, next year, and in 10,000 years. People in ancient civilisations worked out how to predict the position of the sun in the sky, and used this knowledge in the design and alignment of the Egyptian pyramids, Stonehenge and other monuments. This information was clearly viewed as important by ancient people, just as it is critical to our own lives. The path of the sun largely determines how much solar energy will be received by each square metre of the Earth's surface. This determines seasonal variations and is a major factor determining climate.

The Earth spins on its axis, which is tilted relative to the plane in which the Earth orbits around the sun. The "declination angle" of the Earth is 23.5°. This tilt of the Earth's axis is responsible for the seasons throughout the year.

Because of the Earth's daily rotation around its north-south axis, the sun appears to travel within a circular arc around the Earth.

Life in the tropics

At the equator (zero degrees latitude), the sun moves within a circular arc within a plane parallel to the equator. At either of the two equinoxes (when the days and nights have equal length, on or about March 21 and September 21), the sun moves within the plane passing directly through the equator. On these days only, the sun rises exactly due east, passes directly overhead, and sets due west. At other times of the year, this plane is shifted to the north or south. In mid-summer, on the summer solstice (on or about December 21st in the southern hemisphere), the plane of the sun's path passes 23.5° to the south of vertical at mid-day. By mid-winter, the plane of the sun has shifted to its most northerly position, and at mid-day, passes 23.5 degrees north of vertical.

Thus, at the equator, solar radiation is most intense – striking the Earth's surface most closely to perpendicular – during the spring and autumn equinoxes – not during the "summer". In fact, at the equator, the intensity and angle of the sun are the same in summer and winter (with the sun's position at mid-day alternating to the south and north of overhead). There is no distinct "summer" and "winter" seasons (although many equatorial regions alternate between "hot and wet" and "hot and dry" seasons).

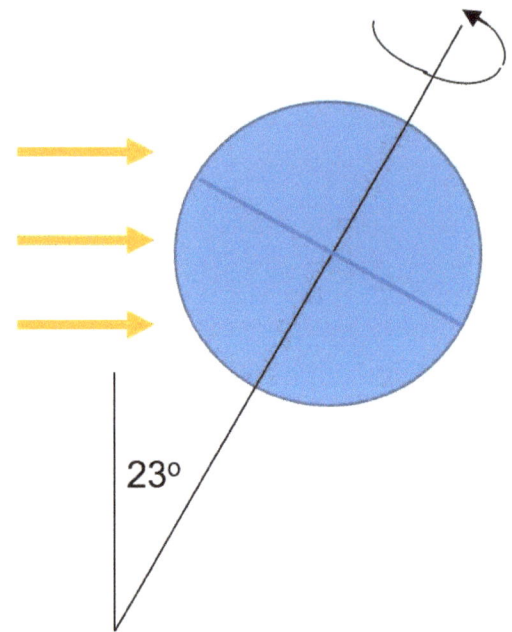

The region of the Earth, within 23.5 degrees of the Equator is referred to as "the Tropics". Applying simple trigonometry to the spherical shape of the Earth, we can show that the Tropics comprises 40% (given by sine 23.5°) of the Earth's surface. If we look at where the sun will be at mid-day at any location within the tropics, it alternates between north and south of directly overhead throughout the year. Anywhere else on Earth, the sun will *always* pass *either* north *or* south of vertically overhead (depending on whether you are in the southern or northern hemisphere respectively). Due to the high solar radiation throughout the year (with rays of the sun coming from near-vertical at mid-day), tropical climates tend to be hot all year.

As we move away from the equator, the plane through which the sun moves begins to tilt. And, the further you go from the equator, the more the plane of the sun is tilted. As it turns out, *wherever you are on Earth*, *the sun moves in a circular arc tilted from a vertical plane at the latitude angle*. At latitudes greater than 23.5°, in the temperate zones, the sun's position at mid-day will *always* pass either to the north of vertical (in the southern hemisphere) or to the south of vertical (in the northern hemisphere). Sunlight shines most directly on the surface in mid-summer, and at the most oblique angle in mid-winter. Consequently, the temperate zones in the northern and southern hemispheres are characterised by distinct summer and winter seasons.

Life near the poles

Let's look at the path of the sun at the south or north pole, at 90 degrees latitude. Here, the sun always travels in a circular arc *parallel to the horizon* (that is, tilted 90° from a vertical plane). In mid-summer, the sun travels in a circular arc at a constant angle of 23.5° above the horizon. This arc passes above the eastern horizon in the "morning", swings above the northern horizon at mid-day, towards the west in the "evening", and then circles back above the southern horizon. Summer is one continuous day, lasting for half the year. However, after mid-summer, the arc of the sun is lower in the sky, and passes closer to the horizon. By the autumn equinox, the sun just skirts along the horizon. After that, the path of the sun goes below the horizon, and continuous darkness follows for the next six months (until the following spring equinox, when the sun again begins to skirt along the horizon).

A similar situation occurs throughout the polar regions. For the southern hemisphere, the polar region extends from the south pole (90 degrees south latitude) to 66.5 degrees south latitude – that is, within 23.5° of the south pole. Throughout this region, the arc of the sun is not exactly parallel to the horizon, but become progressively tilted as we move away from the pole. In mid-summer, the sun is always above the horizon, but never more than 47° (2 X 23.5°) above the horizon at mid-day. A continuous period of daytime extends through the summer, whose duration depends on the latitude angle. In mid-winter, the sun is always below the horizon in a period of continuous night. In-between summer and winter are periods when the sun swings just above the northern horizon in daytime, and just below the southern horizon at night. The situation in the northern polar region (the Arctic) mirrors that of Antarctica in the south polar region. The Arctic and Antarctic polar regions occupy 8.3% of the Earth's surface (given by 1-cosine 23.5°).

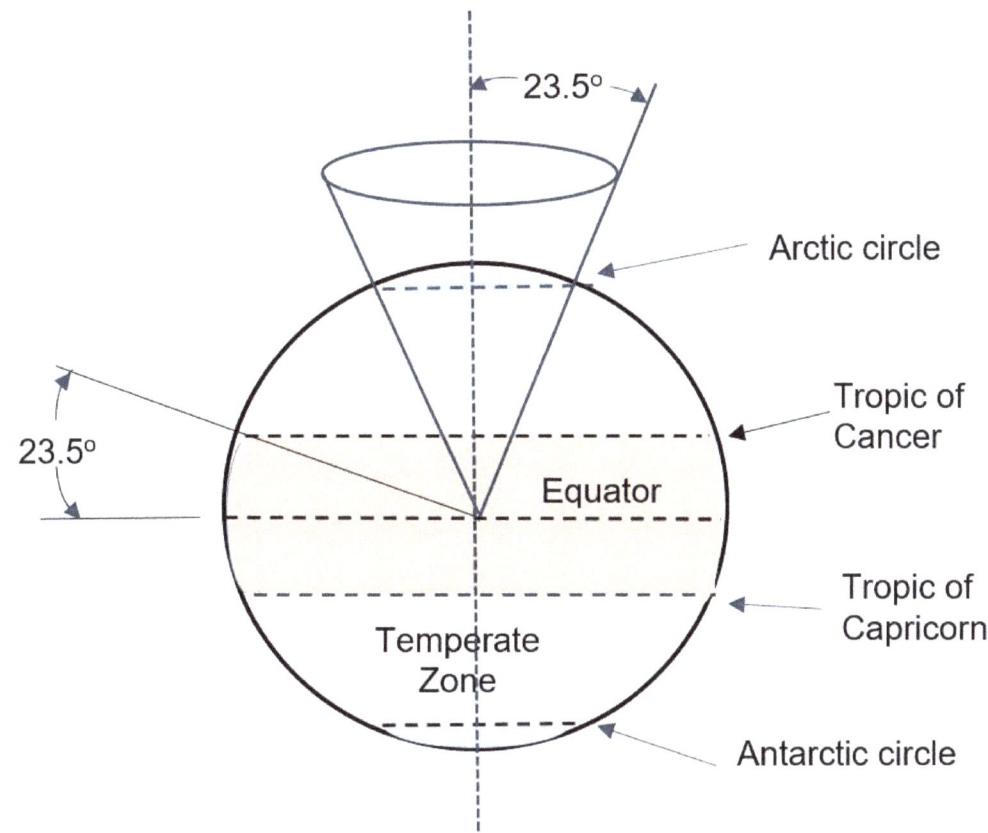

House design to suit local geography and climate

Most of the Earth's population (including those of us in Brisbane) live within the Temperate Zone, which consists of two bands (in the northern and southern hemispheres) lying between 23.5° and 66.5° latitude. Brisbane, located at 27.5° latitude, is just within the temperate zone (and hence is referred to as "Sub-tropical"). The Temperate Zones in the northern and southern hemispheres account for 52%, or just over half, of the Earth's surface.

In Brisbane, the sun passes within a circular arc which is tilted from the vertical at the latitude angle of 27.5°.

- At the spring and autumn equinox, the sun rises exactly in the east, passes 27.5° to the north from vertical at mid-day, and then sets exactly in the west.

- In the summer months, the arc of the sun shifts to the south. At the summer solstice (the longest day of the year, around December 21st), the sun rises in the southeast, passes just 4 degrees to the north of vertical (27.5° - 23.5° = 4°) at mid-day, and sets in the southwest.
- In the winter months, the arc of the sun shifts to the north. At the winter solstice (the shortest day of the year), the sun rises in the northeast, passes 51° (27.5° + 23.5°) north of vertical at mid-day, and sets in the northwest.

Areas in the temperate zone tend to have distinct summer and winter seasons. Dwellings should be designed to provide a comfortable environment for the occupants, reducing the need for air-conditioning on hot summer days and for heating during cold winter periods. One critical aspect of energy-efficient house design is preventing or minimising sunlight entry through windows in summer, while allowing sunlight to enter and warm the building interior in winter. Differing paths of the sun across the sky in summer and winter provide opportunity for windows to selectively allow sunlight to enter in winter, while providing full shading in summer.

In the southern hemisphere, at Brisbane's latitude, solar entry is best controlled by locating the main living and window areas of buildings on the north-facing wall with an awning extending about half the window height. Cross-ventilation through open-plan living areas can then be achieved by locating fully-openable windows on south-facing walls. West-facing windows can be very difficult to shade and problematic in the summer, as these windows receive direct sunlight from low in the sky in late afternoon (usually the hottest time of the day). Ideally, the western side of a house should house service areas such as bathroom, laundry, garage and storage, which are not often occupied and require relatively small windows. However, such design considerations may be difficult to apply, depending on the site orientation, layout and views.

In the southern hemisphere, placing solar panels or solar water heaters on a north-facing roof maximises solar energy collection throughout the year. Maximum solar output is obtained throughout the year by orienting solar panels facing north and tilted at the latitude angle. If maximum output is desired during the summer (to power air-conditioning, for example), solar panels should be tilted at a less steep angle. Conversely, solar power collection can be maximised in winter (with some sacrifice of summer output) by mounting solar panels more steeply angled towards the north. Some panel mounting systems allow the angle of solar collectors to be varied during the year (for those homeowners who don't mind clambering upon on their roof).

21. Water world

The Earth contains an estimated 1.4 billion cubic kilometres of water. Of this, the vast bulk (about 97.5%) is water in the oceans, which has a high salt content. Of the fresh water on Earth, much is locked up in glaciers in the Arctic, Antarctic and Greenland. Most of the remainder is located underground in porous rock and soil. Only about 0.3% is fresh water in lakes, swamps and rivers.

Water is continually cycled between the oceans and fresh water through evaporation and condensation as rain. This plays a major role in the Earth's climate by transferring heat around the planet (from the equatorial to polar regions) and by absorbing heat during the day and releasing heat at night.

Earth is very much a "water world", but the same role could be played by evaporation and condensation of other liquid compounds on other planets.

- In 2004, the Cassini-Huygens spacecraft undertook radar mapping of the surface of Saturn's moon Titan and discovered lakes composed of liquid methane, and signs of erosion and river deltas formed by flowing liquid methane (at temperatures of around -180°C). On Titan, methane appears to serve the same role in redistributing heat, creating the climate and sculpting the surface as the water cycle does on Earth.

- In 2014, scientists studying a brown dwarf (an under-sized star, too small to maintain nuclear fusion in its core, with a temperature of about 1,000 degrees Kelvin at its outer surface), found evidence of clouds containing droplets of iron. They believe that "rain" of molten iron falls through the atmosphere, and then evaporates at lower elevations.

Prisoner inside a water droplet

We had previously seen that we residents of planet Earth are confined to the surface of our planet by gravity. We can only escape the gravitational force of the Earth by gaining sufficient energy, the "escape energy", to overcome Earth's gravitational attraction.

Molecules in a water droplet, or any liquid, face a similar situation. They are attracted to the surrounding molecules by intermolecular forces. Molecules in a liquid can slide past one another and move within the liquid (a bit like a passenger on a crowded train trying to reach the door). After moving **within** the liquid, a molecule will have different neighbours, but will still have the same **number** of neighbouring molecules. So long as it remains within the liquid, a molecule experiences the same attractive forces and has the same average energy.

However, a different situation applies to a molecule that tries to escape from the surface of the liquid into the surrounding air or gas. Such a water molecule must overcome strong forces bindng it to neighbouring water molecules in the liquid. Considerable energy is required for water molecules to overcome these attractive forces and to escape into the surrounding air or gas. About 2,200 kilojoules of heat energy are required to vaporise a litre (or a kilogram) of water. This is called the "heat of vaporisation" of water.

This is a large amount of energy – from the viewpoint of a water molecule in the liquid state peering longingly at the surface and dreaming of a life of freedom in the vast world beyond. Typical water molecules at room temperature are moving with a kinetic energy of about 200 kilojoules per kilogram. How could any water molecule hope to ever get enough energy – more than ten times the kinetic energy of an average water molecule - to escape from the confines of the liquid?

The great escape

But all is not lost! There is a way for water molecules to escape the liquid state.

A drop of water contains about 100 billion billion molecules. These molecules are continually exchanging energy by colliding with one another. During a collision, a water molecule may gain or lose energy from its neighbour. At any one instant, a particular water molecule may have higher-than-average energy, while other molecules have less-than-average. Thus, individual molecules vary widely in energy. Exchange of energy between molecules causes some to have more-than-average while other molecules have less energy than average. At normal room temperature, a tiny proportion of water molecules (about one in ten million) have sufficient energy to escape from the liquid to the gas (vapour).

Of course, only the most energetic water molecules can escape from the liquid into the vapour state, leaving behind molecules that have lost energy during intermolecular collisions. Consequently, the escape of the most energetic molecules from the liquid – which we call "evaporation" – lowers the average energy (and thus, the temperature) of the remaining liquid. Evaporation cools the liquid.

The fraction of molecules with sufficient energy to escape varies exponentially with temperature. Each increase in temperature of about 10°C enables about twice as many molecules to escape into the vapour.

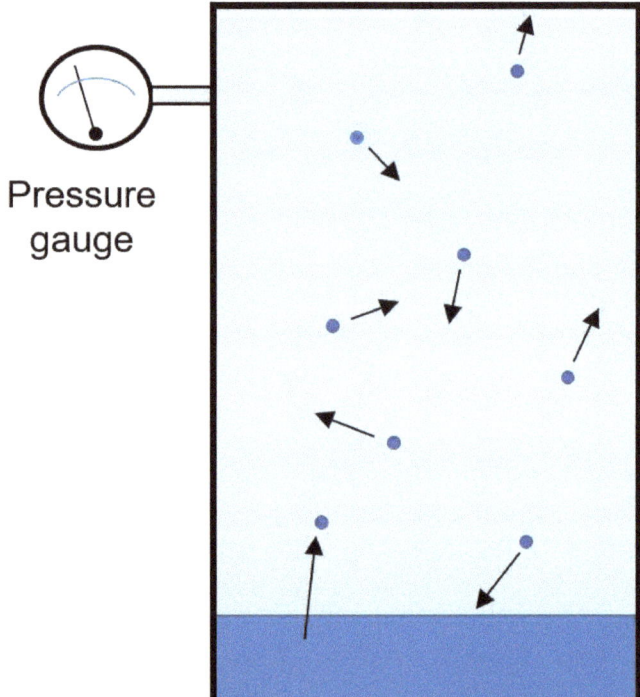

Consider a "thought experiment" in which a sample of liquid water is put into a sealed container which initially contains a vacuum. Initially, no gas or vapour is present in the container, so the pressure inside the container is zero.

A tiny fraction of water molecules have sufficient energy to escape from the liquid and wander freely in the space above the liquid. Over time, more and more molecules are found in the gas phase, and the gas pressure increases.

However, once molecules have escaped into the gas phase, some collide with the surface of the liquid and are re-captured.

A situation is soon reached at which water molecules are escaping from the liquid at the same rate at which water molecules in the gas phase are recaptured by the liquid. Then, the gas pressure remains steady at the "equilibrium vapour pressure".

A graph of the gas pressure versus time would look like this graph:

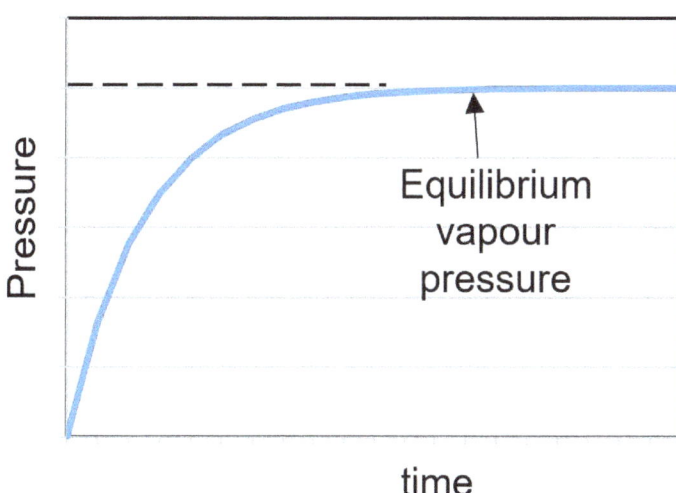

This is a "dynamic equilibrium", as molecules are continually escaping the liquid and other molecules are continually being recaptured. Any particular molecule might find itself in the liquid at 9.00 am, in the vapour at 9.45, back in the liquid at 10.15, etc.

The interesting thing is that for a given liquid (in this case, pure water) the equilibrium vapour pressure **depends only upon the temperature**. For example:

- If we increase the surface area of the liquid, by adding more water to the container or changing its shape, this increases the rate at which molecules escape from the liquid, but also increases the rate at which water molecules in the gas strike the liquid surface and are recaptured. Consequently the equilibrium vapour pressure remains the same.

- If we use a vacuum pump to remove the water vapour, this would initially cause the total gas pressure to drop below the equilibrium vapour pressure of the water. This would reduce the rate at which water molecules return to the liquid from the vapour state, while water molecules would continue to escape at the same rate. The liquid water would boil to restore the equilibrium vapour pressure.

    As we continue to pump away water vapour, and as the water boils to restore the vapour pressure, vaporisation of the liquid absorbs heat and its temperature falls. The liquid becomes colder and it vapour pressure reduces. Eventually all of the liquid water would boil away. At this point, there is no longer liquid water in equilibrium with its vapour, and only then the pressure could fall to zero.

While the equilibrium vapour pressure of a liquid varies only with temperature, the effect of temperature is dramatic. The equilibrium vapour pressure varies exponentially with temperature. For water, increasing the temperature by 10°C causes the equilibrium vapour pressure to roughly double. What that means is that increasing the temperature by 20°C causes the equilibrium vapour pressure to increase about 4 times, increasing the temperature by 30°C causes the equilibrium vapour pressure to increase to roughly 8 times as much, and increasing the temperature by 40°C causes the equilibrium vapour pressure to increase about 16 times. Over the temperature range from 0°C (the freezing point of water) to 100°C (normal boiling point), the equilibrium vapour pressure of water increases nearly two hundred times! This has huge implications for life on Earth, as we shall see.

If we plot the equilibrium vapour pressure of water versus temperature, the graph looks like this:

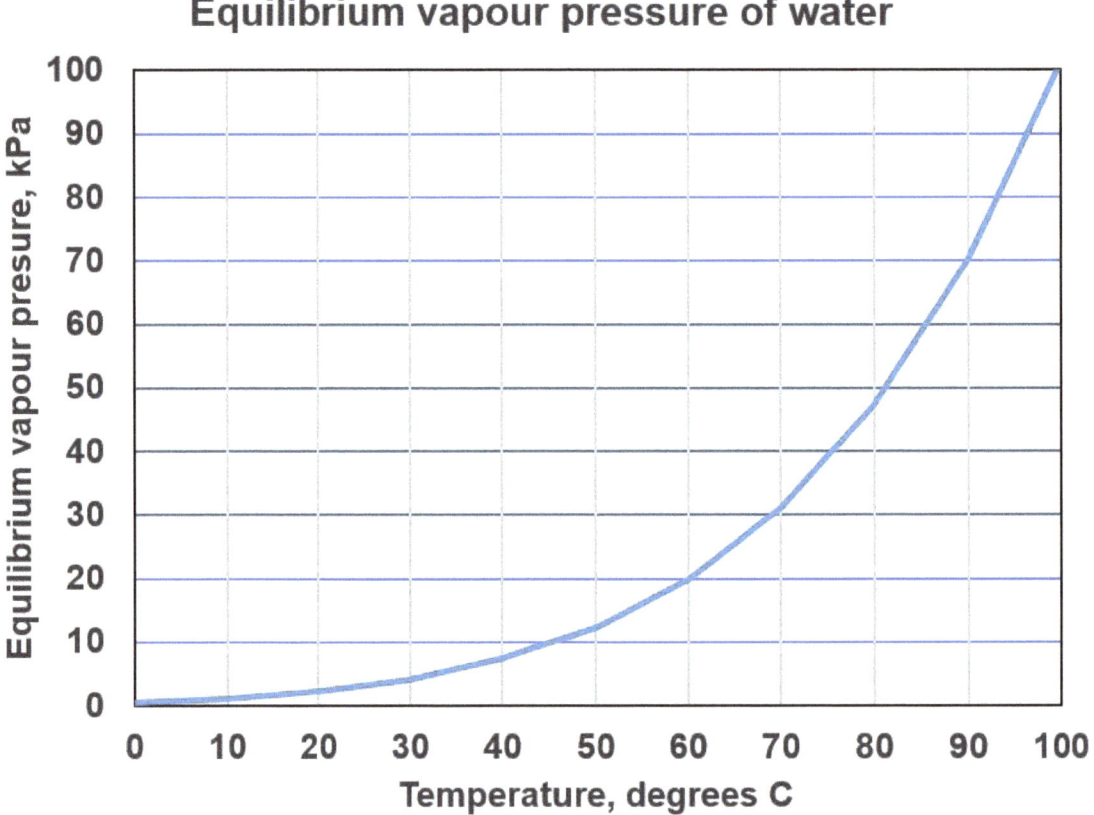

The effect of temperature on vapour pressure is demonstrated quite spectacularly in a well-known high-school experiment. In the "imploding can experiment", an aluminium can of soft drink is emptied of its contents (usually by imbibing the fluid). A small amount (perhaps one or two millilitres of water) is squirted into the can, which is placed over a burner, until the water boils. The water is allowed to boil for a short time until all of the air in the can is displaced by water vapour at 100°C (although, for safety, it is important that the can does not boil dry). At this point, the gas inside the can consists entirely of water vapour at atmospheric pressure (the same pressure acting on the outside of the can).

The can is removed from the heat source with gloves or tongs and then turned upside-down, allowing any remaining hot water to drain out. The can is then placed, with its opening facing downwards into a bowl of cold water. At this point, the hot water vapour (at 100°C) comes into equilibrium with the cold water (at, say, 20°C). Water vapour molecules are captured into the cold liquid as soon as they strike its surface. Nearly all the vapour condenses into the cold liquid, causing the pressure to plummet. The can suddenly implodes, and is pulled out of the experimenter's gloved hand (or tongs).

You can watch this experiment demonstrated in short videos on the internet. Here are a couple of videos that I have found:

http://www.stevespanglerscience.com/lab/experiments/incredible-can-crusher

http://www.youtube.com/watch?v=gl7EhcRuXyQ

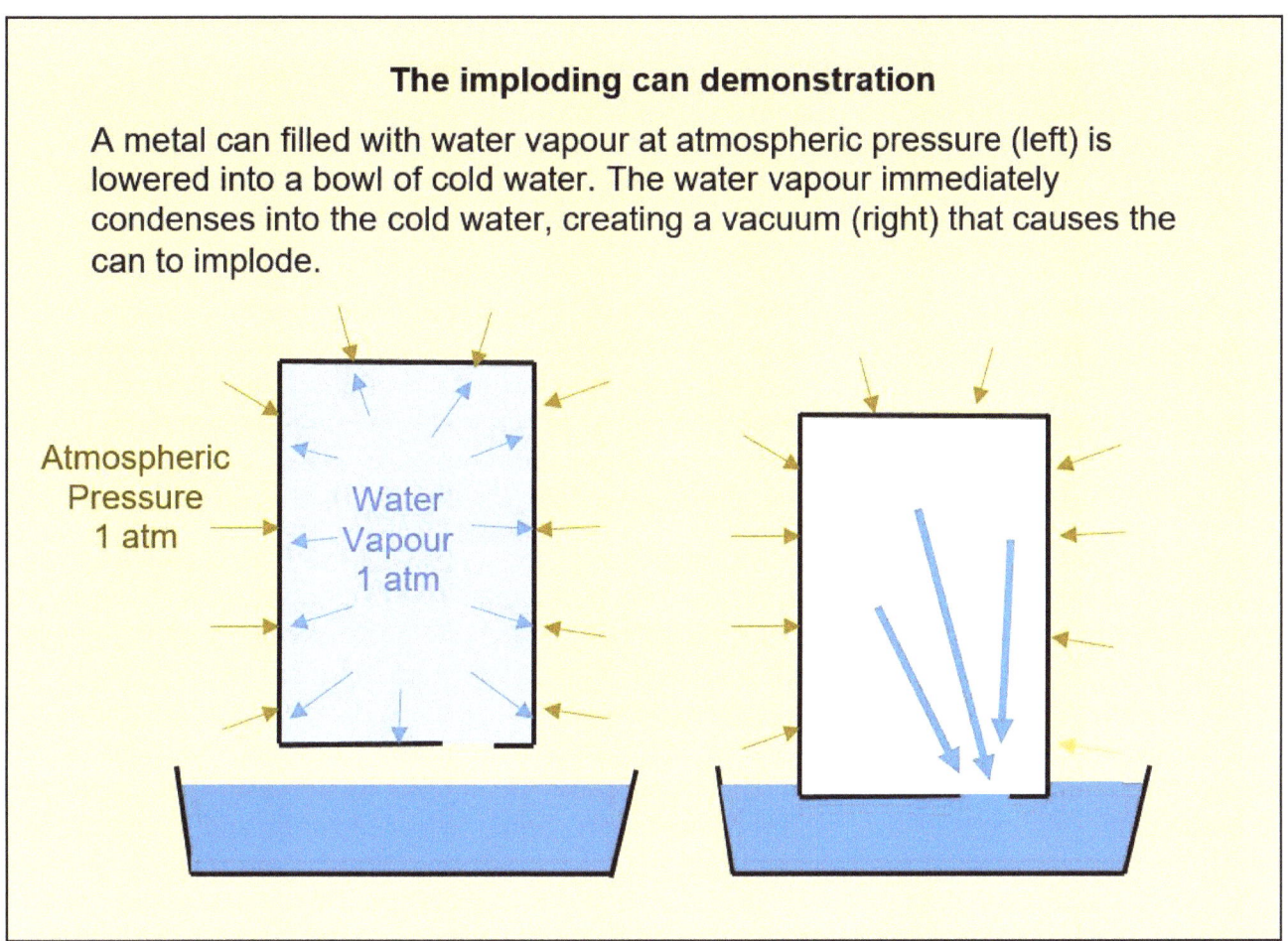

Sceptical viewers may think that the experimenter is cheating by crushing the can in their hand, or dropping the can into the bowl of water. However, I have performed this demonstration many times, and let me assure such readers that the can is literally pulled out of the experimenter's hand and crushed. This demonstration can be done with simple equipment, although if you try it, you should exercise care not to scald or burn yourself (safety glasses are recommended, as with any experiment or demonstration).

The concept of equilibrium vapour pressure is critically important for understanding natural processes occurring on Earth, and has surprising implications. For one thing, many people think that water boils at 100°C, but water can boil at nearly any temperature, depending on the pressure of air or gas. Normally, we experience air pressure of one atmosphere, at which water boils at 100°C. So, the **normal boiling point of water** is 100°C. However, if you lived on the top of Mount Everest, where the air pressure is only about one-third that at sea level, the boiling point of water would only be about 70°C.

The boiling point of water – or any liquid – is determined by the surrounding gas pressure. Whenever the equilibrium vapour pressure exceeds the gas pressure acting on the liquid, then any small bubbles of air or vapour will rapidly expand and rise to the surface. Rising bubbles create turbulence, which causes additional microscopic bubbles of vapour to form, and these bubbles rapidly expand and rise to the surface. Boiling provides a quick and effective means to remove heat by vaporising water. Heat can be provided either by a burner or other **external** heat source, or by the "sensible (**internal**) heat" of the liquid water itself.

The boiling of liquid water at low pressure is shown in the following videos:
> Boiling water at low pressure, Richard Hammond, 2-1/2 minutes
> www.youtube.com/watch?v=XoOQNwcrDWE
> Boiling water in a syringe (no need for a vacuum pump!), 4 minutes
> www.youtube.com/watch?v=I5mkf066p-U

If we placed water at room temperature in a vacuum chamber and then began pumping out the air, the air pressure would decrease until the liquid begins to boil. As water molecules escape into the gas (which is then pumped away), this absorbs heat, and the liquid cools. The equilibrium vapour pressure falls as the liquid cools, so the vacuum pump must reduce the air pressure even further to keep the liquid boiling. The liquid continues to boil, gets colder and colder, until eventually the temperature of the water reaches 0°C and it freezes.

This may sound extremely bizarre and incredible, but when you see it with your own eyes many times, it begins to make perfect sense. This behaviour is not limited to water: exactly the same applies to any liquid. When I was working in an organic chemistry laboratory many years ago, I was routinely working with dilute solutions of chemical products dissolved in a liquid solvent (commonly, solvents such as acetone, ether or toluene are used to dissolve organic compounds). To remove the solvent, leaving crystals of the chemical product, chemists use a piece of equipment called a "rotovap". This evaporates the solvent by applying a vacuum (at the same time, rotating the flask, so the liquid doesn't begin to boil violently and fly out of the flask). The chemistry lab where I worked was not air-conditioned, and on warm humid summer days, I would notice droplets of water condensing and dripping from the outside of the cold flask. As the solvent continued to boil, the moisture on the outside of the flask would freeze into ice. I was intrigued the first time that I saw this, but soon accepted it as perfectly "normal" behaviour for liquids. And, of course, it is.

You can see this yourself by watching the following short videos:
> Water boils, and then freezes, under vacuum, 2 minutes:
> www.youtube.com/watch?v=pOYgdQp4euc
> Boiling/freezing acetone in a flask, 5 minutes
> www.youtube.com/watch?v=oSMiec0bECw

Once you understand how a liquid can be made to boil – and be cooled – by reducing its pressure, it is a small step to understand how a refrigerator or air-conditioner works, how heat engines work, why humans need a pressurised spacecraft or spacesuit to survive at high altitude, and why growing crops consumes most of the fresh water produced on Earth.

22. Keeping cool:
The science of refrigeration and air-conditioning

Perhaps no invention has played a greater role in transforming our society and economy than the refrigerator. Until the advent of large industrial refrigeration plants in the mid-19th century, fresh food could only be transported and distributed over short distances and times. Production of fruit, vegetables, dairy products and meat had to be undertaken within short distances of towns and cities. But the nexus between agricultural production and population centres was broken by refrigeration and freezing. It enabled the agricultural sector and cities to develop largely independently, and catalysed the amalgamation of small family farms into huge, highly mechanised operations, with a massive shift of population from rural to urban areas.

This was perhaps the greatest social revolution of modern history, and refrigeration played a major part, especially in the latter stages. Only several hundred years ago, the vast majority of people lived on farms or in small agricultural communities. During the early history of the United States, about 95% of the population was involved in farming, and 5% lived in towns and cities. Within three hundred years, this completely reversed. This trend is still continuing in Asia and elsewhere as family-owned farms are amalgamated and mechanised, and as rural populations shift into cities.

The introduction of the first household refrigerators, starting around 1914 in the United States, would ultimately change the way people lived in cities and surrounding suburbs. No longer would people need to shop each day at their local baker, butcher and other small shopkeepers. As late as 1960, as I was growing up in the Lower East Side of New York City, I recall a bustling world on Avenue B of small shopkeepers, bakeries, butcher shops, a delicatessen and wizened-looking ladies and men selling products from carts or pickles from a barrel (although there was also a small supermarket on Avenue D). But then, with most families owning a refrigerator and growing numbers owning a car, householders (usually housewives) starting to do weekly grocery shopping at supermarkets and shopping plazas. This, combined with many other factors, enabled women to join the workforce and changed the social fabric of society to what it is today – in many ways for the better.

So how do refrigerators work?

Operation of refrigeration and air-conditioning systems is based on the evaporation/boiling of volatile liquids (refrigerants). As we have seen in the previous chapter, a liquid can be made to boil at nearly *any* temperature by reducing the pressure. Furthermore, we have seen that the process of evaporation (or boiling) absorbs heat. If we cause a liquid to boil by reducing its pressure, the heat of vaporisation is provided by internal heat within the liquid itself, causing its temperature to fall.

Using these simple concepts, let's see how we might design a refrigerator. First, we need to choose a suitable liquid that we could boil by applying a vacuum. We could use water. It is cheap, non-toxic, non-flammable and widely available, but it has two major shortcomings as a refrigerant. Firstly, at the temperature within a refrigerator (0-6°C), the equilibrium vapour pressure of water is only about 1/100th atmospheric pressure. We would need a very large vacuum pump to suck sufficient volumes of water vapour at such low pressure. Secondly, we want to produce temperatures just above 0°C (or below 0°C for a freezer), so a water refrigerant would likely freeze.

The ideal refrigerant would be more volatile than water (that is, have a normal boiling point well below that of water) and would have a lower freezing point. Of course, the ideal refrigerant should also be cheap, non-toxic, non-flammable, environmentally benign and readily available. Unfortunately, no known refrigerant fully meets all these requirements.

Early industrial refrigeration plants used liquid ammonia as refrigerant. With a normal boiling point of -33°C, ammonia has sufficiently high vapour pressure and other properties to be a practical refrigerant. However, ammonia is toxic at high concentrations, and fatal accidents involving leaks at commercial refrigeration plants caused ammonia to fall out of favour. Other compounds used as refrigerants, such as methyl chloride and sulfur dioxide, were also highly toxic.

In the 1950s and 1960s, a series of synthetic chemical compounds called "freons" became widely adopted for refrigeration. The most common was Freon-12 (dichlorodifluoromethane, normal boiling point = -30°C). These compounds seemed to be safe and ideally suited for as refrigerants. They are extremely chemically stable, and so, don't burn and don't decompose in the atmosphere. Years later, researchers discovered that these compounds eventually diffuse to the upper atmosphere, where they release chlorine free radicals. These radicals destroy ozone molecules that provide a protective shield that absorb ultraviolet light in sunlight. The ozone layer in the upper atmosphere performs a critical role protecting humans, as well as all other animals and plants living on the surface of the Earth. Once the threat of ozone destruction was recognised, the production and use of freon compounds was phased out and prohibited through the 1987 Montreal Protocol. This was one of the very few highly effective international collaborations to protect the environment.

The class of compounds that was adopted to provide a direct substitute for freons in the short-to-medium term was fluorinated hydrocarbons, such as HFC 134a (1,1,1,2 – tetrafluoroethane, normal boiling point = -26°C). These compounds have a very low ozone depletion effect, but are very powerful greenhouse gases. For example, HFC 134a has a "Global Warming Potential" that is 1,430 times that of carbon dioxide.

Eventually, fluorinated hydrocarbons like HFC 134a are expected to be phased out. Some refrigeration and air-conditioning systems use hydrocarbons, such as propane (normal boiling point = -42°C) or butane (normal boiling point = -1°C). Hydrocarbon compounds pose a fire risk in the event of a leak, although the risk can be minimised by incorporating relatively small quantities of refrigerant within the system. Eventually, most refrigerators and air-conditioners are likely to use hydrocarbon refrigerants.

Let's say that we choose pentane as a refrigerant. Pentane is a liquid at room temperature but, with a normal boiling point of 36°C, readily boils when the pressure is reduced just below atmospheric pressure. The freezing point of pentane is -130°C, which is well below the temperatures produced in a refrigerator or air-conditioner, so freezing is not an issue.

To make a refrigerator, we could simply mount a container of pentane within an insulated refrigerator cabinet, and use a vacuum pump to reduce the pressure of pentane vapour, causing it to boil. As the pentane boils, it absorbs heat from (and chills) the vegetables, cheese and milk inside the refrigerator cabinet. The container of boiling refrigerant is called an "evaporator". It is normally made of metal tubing and has fins to increase the surface area for heat transfer (and some refrigerators have a small fan to circulate cold air within the fridge).

So, the boiling pentane cools inside the fridge, but this is only half the story. What happens to the pentane vapour that is sucked from the evaporator? We don't want to discard this into the surrounding atmosphere, and then have to continually add more liquid pentane to replace the liquid that boils away.

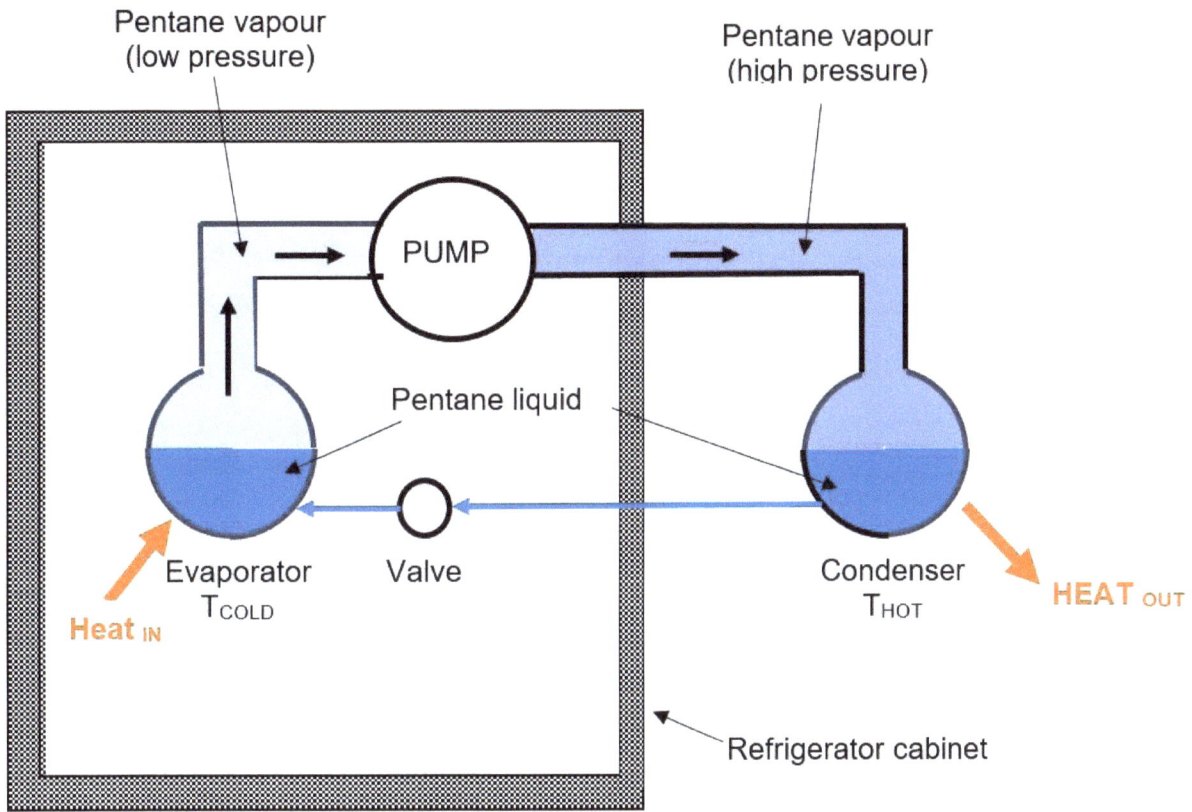

We collect the pentane vapour in another container mounted **outside** the refrigerator cabinet, and we force the pentane to condense back into liquid by increasing its pressure above the equilibrium vapour pressure. Just as we can cause a liquid to boil at low temperature by **reducing** the pressure, we can cause vapour to condense into liquid by **increasing** its pressure. The container used to condense refrigerant vapour is also commonly made of metal tubes with fins, and (as you might expect) is called a "condenser".

The heart of the refrigerator is the pump, which (literally) does all the work. Basically, the pump causes pentane liquid to boil at low temperature T_{COLD} (say, 5°C inside the fridge), and causes pentane vapour to condense back to liquid at higher temperature T_{HOT} (say, 25°C within our kitchen).

Since the condenser is mounted outside the refrigerator, where the temperature is higher, *the pressure inside the condenser is higher than in the evaporator*.

The pump does work in sucking pentane vapour from the

evaporator at low pressure, and then compressing and pushing it into the condenser at higher pressure. In other words, the pump works against a pressure difference **ΔP** equal to the difference in the equilibrium vapour pressure of the refrigerant at T_{HOT} and T_{COLD}.

Eventually, liquid pentane within the evaporator would boil away, and liquid pentane would collect in the condenser. To avoid this, liquid refrigerant is continually recycled from the condenser to the evaporator through an automatic control valve. No pump is needed to recycle the liquid, since the condenser is at higher pressure than the evaporator.

Pentane refrigerant cycles continually from the liquid state to vapour (inside the fridge at low temperature and low pressure), and then from vapour back to liquid (at high temperature and pressure). The compressor maintains the required pressure difference **ΔP** for this to occur.

Energy required for refrigeration

A refrigerator or air-conditioner is a "Heat pump". It transfer heat from a cold "Heat Source" (inside the fridge) to a hot "Heat Sink" (the surrounding air at higher temperature). Effectively, a refrigerator pumps heat "uphill". To do so, it must consume energy.

The Second Law of Thermodynamics states that heat cannot be transferred from a cooler to a hotter body *unless work is done*. We all know this intuitively. We have all seen that a hot cup of tea will cool to room temperature all by itself, but we also know that a warm bottle of beer will not get cold if we leave it on the kitchen counter. If we want to chill a warm bottle of beer, we must do work – or rather, the pump inside our refrigerator must do work. So, how much work is required to transfer each Joule of heat energy?

Let's assume that we have built the "perfect" refrigerator – one that has no friction, no losses, and perfect heat transfer from the evaporator and condenser. In fact, let's assume that we have built the most efficient refrigerator that is allowed by the Second Law of Thermodynamics. Let's consider how much work must be done to transfer an amount of heat Q_{COLD} from inside the refrigerator at temperature T_{COLD} to the surrounding air at temperature T_{HOT}.

As it turns out, using a bit of mathematics, we can readily calculate the work required to operate such a perfect refrigerator. Firstly, we start with the exponential relationship between the vapour pressure of the refrigerant and its temperature. We assume that the refrigerant vapour acts like an "ideal gas" (that is, that the pressure is directly proportional to the amount of gas per unit volume and its absolute temperature). We assume that the evaporator and condenser have very large surface areas that offer no barrier to heat transfer. We also assume that no heat is transferred to the pump as the refrigerant vapour is compressed. Finally, we assume that the temperature difference between the heat source and heat sink ($T_{HOT} - T_{COLD}$) is small, much less than the absolute temperature T_{COLD}. These assumptions greatly simply the calculation, allowing us to derive an astonishingly simple result for the minimum work required to transfer an amount of heat Q_{COLD} from inside the fridge.

$$\text{Minimum Work required} = Q_{COLD} \left[\frac{T_{HOT} - T_{COLD}}{T_{COLD}} \right]$$

Note that the temperatures T_{HOT} and T_{COLD} are *absolute temperatures* (that is, relative to absolute zero, -273°C).

For a refrigerator or air-conditioner, we want to **_transfer the maximum amount of heat_** from the heat source with the **_minimum work input_**. We define the "Coefficient of Performance" (COP) of a refrigerator or heat pump as the ratio of Heat transferred divided by the work input. From the point of view of energy efficiency, we want the Coefficient of Performance to be as high as possible so that we use the minimum energy to chill our beer and keep our vegetables cold.

The maximum Coefficient of Performance that could be achieved by an ideal, perfect refrigerator (or any Heat Pump) is simply:

$$\text{Maximum Coefficient of Performance} = \left[\frac{T_{COLD}}{T_{HOT} - T_{COLD}} \right]$$

Bear in mind that the COP of a refrigerator or an air-conditioner only tells part of the story. We can have a refrigerator with a high COP which consumes a lot of energy because heat enters the refrigerator cabinet through poorly insulated walls or leaking door seals (so Q_{COLD} is large). Similarly, an efficient air-conditioner can use excessive energy if lots of heat enters the building through unshaded windows and uninsulated roof and walls.

If we are air-conditioning a house, where the inside temperature is 23°C (296° absolute) and the outside temperature is 10°C hotter than inside, then it would theoretically be possible to achieve a COP as high as 30. In reality, with current air-conditioning technology, we would never get anything like a COP of 30. A typical room air-conditioner might achieve a COP of 3-4. A listing of actual measured COPs for all air-conditioner models sold in Australia is available on the Energy Rating website operated by the Australian Government (www.energyrating.gov.au).

But even a COP of 3 is still a bargain. Look at it this way: we only need to put in one unit of work energy to remove 3 units of unwanted heat energy. It looks like we are cheating the Laws of Thermodynamics, but we are not. A refrigerator does not get rid of heat, it effectively pushes heat "uphill" to a slightly higher temperature, from around 2°C to 25°C (275° absolute to 298° absolute), so that it can flow into the heat sink.

The consequences of the 2nd Law of Thermodynamics are clear: To achieve the highest efficiency (COP) for any heat pump, the temperature difference against which the heat pump is operating, $T_{HOT} - T_{COLD}$, should be as small as possible.

Other applications of heat pumps

Even in applications involving moderately high temperature differences, heat pumps can still provide significant reductions in energy consumption. For example, heat pumps provide an effective alternative to solar water heating by using the air as a heat source. In Queensland, daytime temperatures are typically in the range 20° – 30°C throughout the year, which is warm enough to serve as a heat source for water heating.

Water heating accounts for a major share (about 30%) of the total energy used in a typical Queensland household, as well as much of the energy used in restaurants, laundries and many other commercial and industrial operations.

In the past, many households used hot water storage tanks heated by electric resistance heating, and this was widely encouraged by the availability of cheap off-peak electricity tariffs (which allowed water to be cheaply heated during night-time hours, and stored for daytime use). Electric resistance heating is wasteful (from a thermodynamic viewpoint), since it converts a high-grade energy source (electricity) into low-grade energy (heat at about 60°C).

Other possible energy sources for water heating include natural gas and solar energy.

Another option, which is very suitable where there is limited access to sunlight or low-cost natural gas, is a heat pump water heater. The heat pump is basically a refrigerator – but operated for a different purpose. It absorbs heat from the surrounding air (typically at 20°C), and "upgrades" and pumps this low-grade ambient heat into a water tank at higher temperature (typically 60°C).

The maximum possible Coefficient of Performance that could be achieved by a heat pump with a heat source of 20°C (293° absolute) and a heat sink of 60°C is given as follows:

$$\text{Maximum Coefficient of Performance} = \frac{T_{COLD}}{T_{HOT} - T_{COLD}} = \frac{293}{60-20} = 7.3$$

Typical heat pump water heaters have a COP of about 3, which means that they use only one-third as much electrical energy as a conventional electric water heater - while providing the same amount of hot water at the same temperature. So, if the source of electricity is coal-fired power stations, only one-third as much coal must be burned, releasing one-third the amount of greenhouse gases and other pollution (nitrogen oxides, sulfur oxides, particulates, etc). That's a huge advantage from an environmental viewpoint, and a huge benefit to the local and global community. However, from an economic viewpoint, the advantage to the end user is not so clear-cut. Firstly, heat pump water heaters are significantly more expensive than conventional electric resistance water heaters. To be economically competitive, energy cost savings during the lifetime of a heat pump water heater should offset the higher purchase cost of the unit. Secondly, many heat pump water heaters do not have sufficient capacity to operate only during night-time hours, and therefore, cannot take advantage of cheap off-peak electricity tariffs (available only at night).

So, depending upon electricity tariffs and other factors, the "life-cycle cost" of heat pump water heaters tend to be roughly comparable to those of conventional electric water heaters, even though they use much less electricity. Until recent years, there was no strong *economic* incentive for a householder to buy a heat pump water heater instead of an electric resistance water heater. This represented a "market failure", or a dis-connect between what is in the best interests of the community (human society) and the economic interests of the person who is purchasing a water heater. Within the past decade, however, higher electricity tariffs and government regulations have shifted the balance away from conventional electric water heaters.

By the way, heat pumps have also been applied to clothes driers, another type of energy-intensive appliance. Conventional clothes driers use electric resistance heaters to heat air circulating inside the drier to evaporate water.

Heat pump clothes driers are sold in Europe, but apparently, are not available for sale in Australia (presumably because consumers would not be prepared to pay two or three times the price for a clothes drier that is more efficient). Heat pump clothes driers have a Coefficient of Performance of about 2, and thus, use half the electricity of conventional clothes driers. In Europe, where energy tariffs have always been higher than in Australia (and where many countries now rely on Russia for their gas supply), there has long been a much stronger push for energy-efficiency than in Australia.

In Australia, some people use a clothes drying technology which is even more efficient than heat pump driers – hanging clothes on a line (such as the iconic Hills Hoist), which relies entirely on renewable solar energy. Of course, this venerable technology has its own limitations, which are becoming more pronounced as people increasingly live in high-rise units and small house allotments, and increasingly expect to be able to clean/dry their laundry whenever they feel like it (including rainy/cloudy days).

The case of heat pumps illustrates some important general points relating to energy-efficiency. Technology is obviously a critical part of the solution to reduce the world's energy and greenhouse challenges, but it is only one component. More efficient technology has generally been adopted only if it is economically competitive, or if governments intervene in the market (and governments are loath to impose restrictions or costs on their local constituencies, when the benefits will accrue to the entire world at some later time). Furthermore, technology is only accepted and adopted if it is consistent with people's expectations, desires and culture.

22. Heat engines and the limits of power production

A refrigerator or air-conditioner is a "Heat Pump". It transfers heat from a Heat Source to a Heat Sink at **higher** temperature (effectively, it pumps heat "uphill"). We can use very similar technology to do exactly the opposite – to PRODUCE work by taking in heat at high temperature (T_{HOT}), and allowing it to flow "downhill" to the temperature of the surrounding environment (T_{COLD}). In the process of "flowing downhill", some heat is converted into useful work, and some must be dumped to the atmosphere (the heat sink). This is the idea of a HEAT ENGINE. In fact, an Organic Rankine Cycle engine (or a closed-cycle steam engine) has the same components as a refrigerator, and works on exactly the same principle – but in reverse. This time, the pressure of the volatile liquid in the evaporator (now called a "boiler" in steam engine terminology) is greater than the pressure in the condenser, so the vapour does work as it expands in a piston pump or turbine.

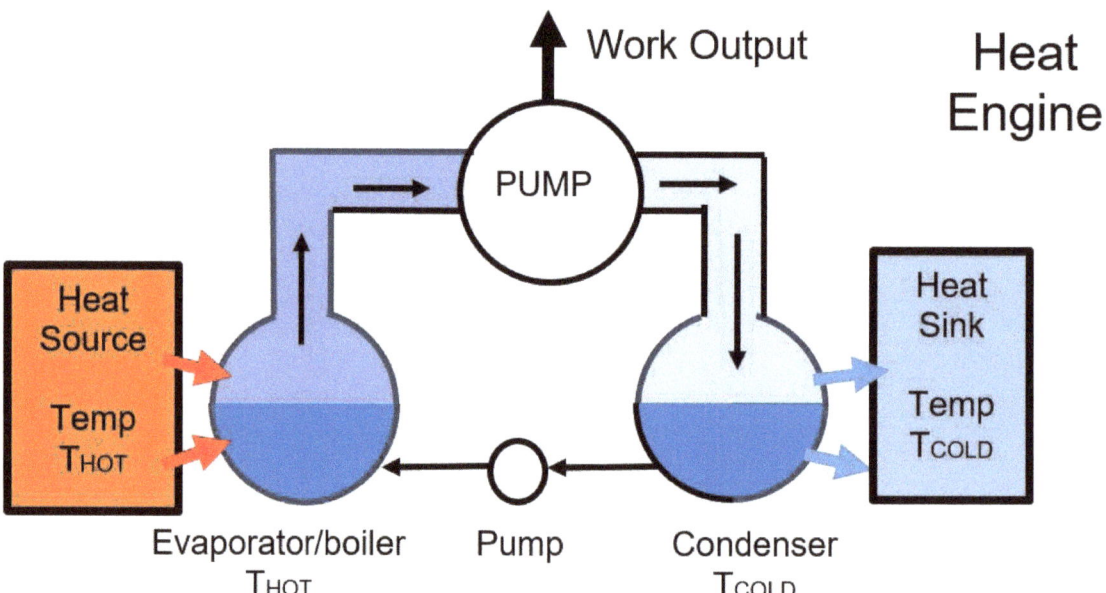

We have seen that the Second Law of Thermodynamics sets a limit for the *minimum work* required to transfer heat "uphill" from temperature T_{COLD} to high temperature T_{HOT}. In the same way, the Second Law of Thermodynamics sets a limit for the *maximum work* that can be produced when heat is transferred "downhill" from a high temperature heat source T_{HOT} to the surrounding environment at temperature T_{COLD}.

We can determine the maximum efficiency of a heat engine by conducting a "thought experiment". Imagine a situation where we have an ideal HEAT PUMP (achieving the maximum possible Coefficient-of-Performance) transferring heat from a body at temperature T_{COLD} to a body at temperature T_{HOT}, and then we recycle the heat through an ideal HEAT ENGINE, taking heat from the hot body, producing work, and then dumping waste heat in the cold body.

Imagine that heat is continually cycled from the heat source through the heat engine (where it produces work going "downhill") and then is dumped into the heat sink. At the same time, imagine that heat is absorbed from the heat sink, pushed "uphill" by the heat pump (requiring work to be done), and then pushed into the heat source.

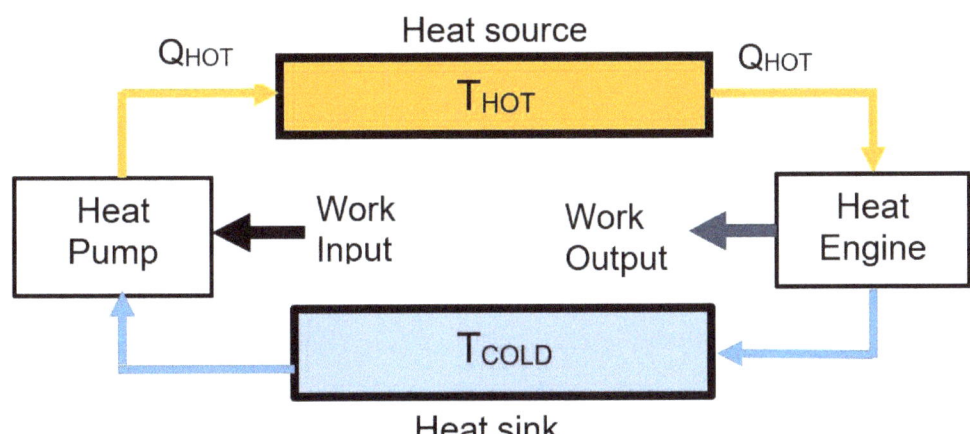

It should be evident that, if our hypothetical apparatus is not to violate the Second Law of Thermodynamics, the work output produced by taking heat Q_{HOT} from the heat source and putting it through the heat engine cannot be greater than the work required for the heat pump to push heat Q_{HOT} back into the heat source. In other words, even a "perfect" heat engine (the most efficient allowed by the Second Law of Thermodynamcs) can only convert a fraction of heat energy Q_{HOT} into work. The remaining heat input must be dumped into the heat sink. The maximum efficiency that an ideal heat engine could possibly achieve can readily be shown to be:

$$\text{Maximum efficiency of heat engine} = \frac{\text{Work output}}{\text{Heat Input } Q_{HOT}} = \frac{T_{HOT} - T_{COLD}}{T_{HOT}}$$

And the maximum work output of a heat engine = $Q_{HOT} \left[\dfrac{T_{HOT} - T_{COLD}}{T_{HOT}} \right]$

Achieving maximum efficiency for a heat engine requires the opposite condition as applies for a heat pump – namely, the temperature difference $T_{HOT} - T_{COLD}$ should be as *large* as possible. Since T_{COLD} is usually the temperature of the surrounding atmosphere, which we cannot control, it is critically important that heat engines operate with a heat source at the highest possible temperature.

Note that the maximum possible Coefficient-of-Performance for a heat pump, and the maximum possible efficiency of heat engine, are **absolute**. It does not matter what technology is used, or how cleverly it is designed. It does not matter if, within the next 200 years, scientists discover entirely new technologies for heat engines. The efficiency of a heat engine will **never** exceed $(T_{HOT} - T_{COLD})/T_{HOT}$ - unless someone figures out a way to violate the Second Law of Thermodynamics (and I would give 10:1 odds that this will never happen). This does not mean that there is no scope for improving the efficiency of heat engines and heat pumps. Even with the best technology that we have now, most heat engines achieve about half of the maximum possible efficiency.

With current technology, the efficiency of heat engines is limited mainly by the maximum temperature that can be withstood by the materials used in the engine. Most electricity generating plants operate on a steam cycle. Water is boiled into high pressure steam within a boiler, and the steam is expanded in a steam turbine. The heat source is generally provided by combustion of coal, or by a nuclear reaction. Coal and other fossil fuels typically produce flame temperatures of 1,500°C or more, but steam cycle power plants operate with a steam temperature of about 500°C. At higher temperatures, steel tubes used to contain the high-pressure steam begin to lose strength. To avoid the possibility of the steam tubes bursting (causing a violent explosion and requiring an immediate shut-down and repair of the plant), the steam temperature is generally limited to about 500°C.

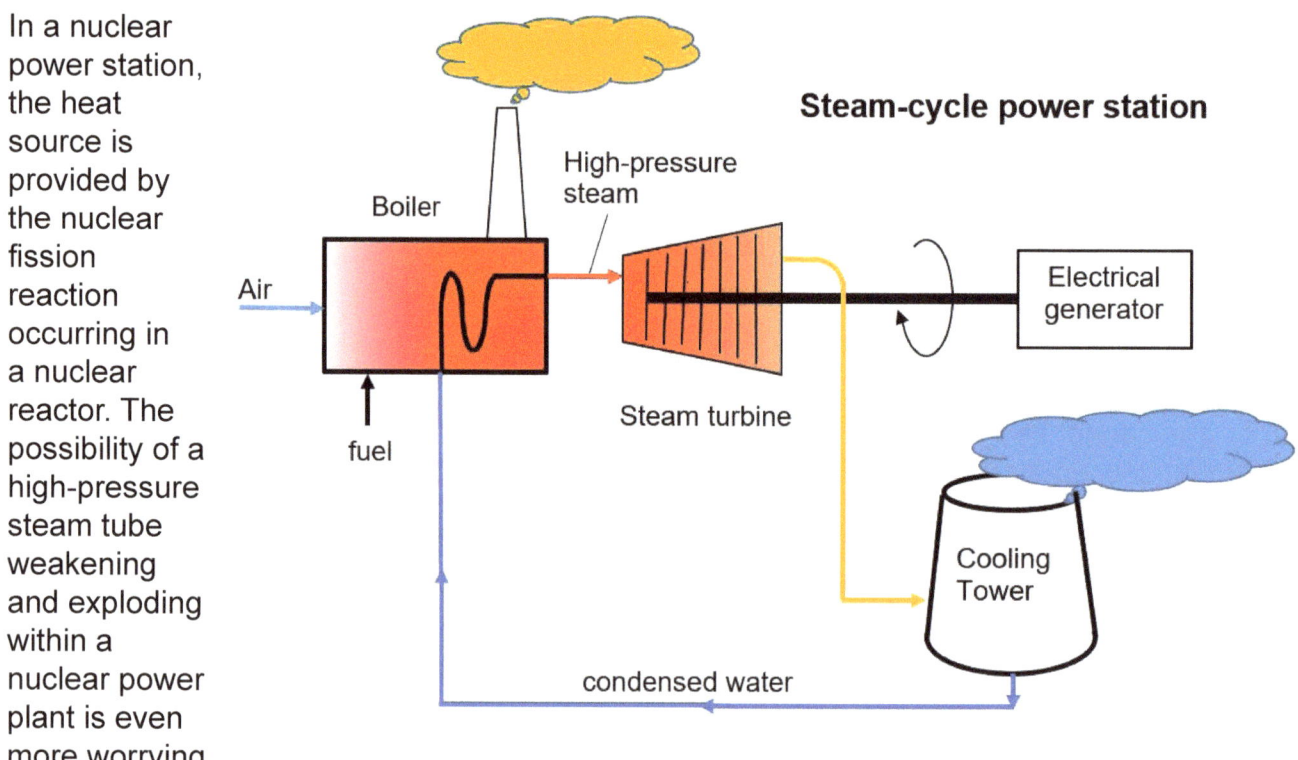

In a nuclear power station, the heat source is provided by the nuclear fission reaction occurring in a nuclear reactor. The possibility of a high-pressure steam tube weakening and exploding within a nuclear power plant is even more worrying than in a coal-fired power station, so nuclear power plants are generally designed to operate at slightly lower steam temperature (with a correspondingly slightly lower efficiency).

The maximum possible efficiency of a power station with a steam temperature of 500°C (773° above absolute zero) and air temperature of, say 27°C (300° absolute), is (773-300)/773 = 61%. The best that we could ever hope for would be to convert 61% of the energy released by the burning fuel into useful work output (mechanical power or electricity) The remaining heat must be dissipated to the atmosphere (usually in a cooling tower).

In actual practice, coal-fired power plants achieve efficiencies of about 35%, slightly more than half the theoretical maximum. Several factors account for additional losses. The boiler cannot recover 100% of the heat released by the burning fuel, so exhaust gases carry residual heat up the stack to be discarded to the atmosphere. Additional losses occur in the steam turbine and in the generators which convert mechanical power produced by the steam turbine into electricity. Finally, losses occur in condensing the steam back into water, which occurs above ambient air temperature.

Such steam-cycle plants normally operate by burning coal or using a nuclear reactor as a heat source. These energy sources are generally much cheaper than natural gas or other fuel sources, but steam-cycle power stations are generally very large (typically, with units of around 1,000 megawatts), and very costly to build. Because of economics and technological limitations, steam-cycle plants are normally operated as "base load" power plants – operating day and night, weekdays and weekends – as much of the time as possible.

Utilities that own and operate coal-fired or nuclear steam-cycle plants want to keep them operating and producing electricity (and earning revenue) continuously at their rated power output. Since these plants are expensive to build, their owners must pay high financing costs – whether or not the plant is generating electricity. Coal and nuclear fuel are generally cheaper than other fuels, so these plants can operate profitably even if they sell the electricity at discounted off-peak rates. Also, the technology of steam-cycle plants favours base-load generation. To bring a massive boiler to operating temperature takes many hours, as rapid heating would cause uneven expansion and crack the steam tubes. Similarly, when the output of a steam-cycle plant is not needed, the boiler must be kept hot or allowed to cool very slowly. This means that steam-cycle plants cannot be quickly brought on-line, or quickly taken off-line, to meet short surges in power demand.

Consequently, while steam-cycle generating plants are often the most cost-effective option to meet the constant base load of an electricity grid, additional generating capacity is required to meet short-term variations in power demand.

One type of "peak load generating plant" uses gas turbines. Basically, gas turbines are a modified version of jet engines used in commercial aircraft. The main difference is that gas turbines produce mechanical power output in their central rotating shaft, while the power output of aircraft jet engines is used to accelerate the exhaust gases to high velocity (which produces forward thrust). Gas turbine power stations can be started and shut-down quickly, and are relatively cheap to build. However, gas turbine plants are limited to using relatively expensive fuels – usually natural gas or diesel fuel, as these don't form ash which would damage turbine blades. Because they use expensive fuels, gas turbine generators are normally operated only during peak demand periods, when a premium rate is paid for electricity sold to the grid.

Burning of natural gas or diesel fuel within a gas turbine produces temperatures in excess of 1,500°C, but here too, engine operation is limited to the temperature range at which component materials retain their strength. In a gas turbine, air is compressed by rapidly-spinning turbine blades, mixed with fuel and burned in combustion chambers, and then hot exhaust gases are expanded through spinning turbine blades. The very high rotational speed of the turbine imposes high stress on the blades, which must have very high tensile strength to resist being torn apart. Although special metal alloys are used for the turbine blades, which are air-cooled, temperatures within the engine must be kept within the range that the blades can withstand. As a result, much of the heat of the burning fuel is retained in the exhaust gases, which exit the gas turbine at temperatures above 600°C. Consequently, the efficiency of such "open-cycle" gas turbines is generally about the same (or slightly less) than a steam-cycle plant.

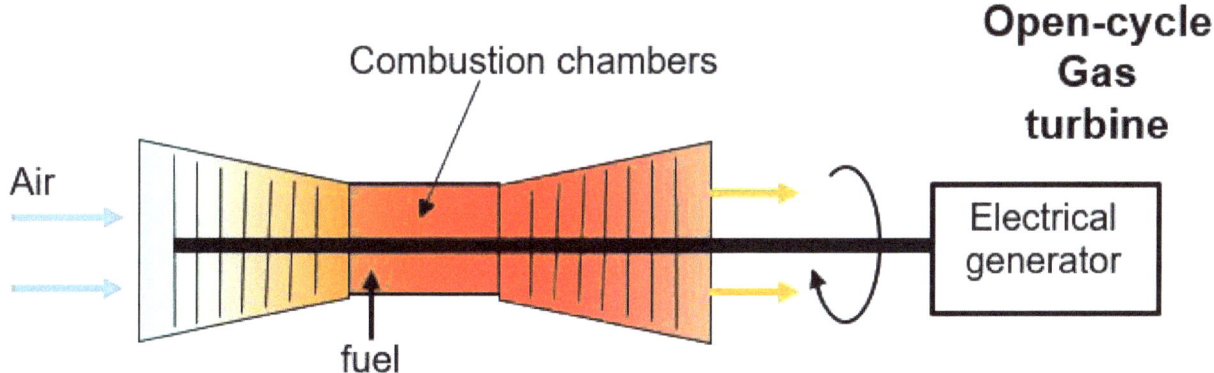

However, the "waste heat" in the exhaust gases of a gas turbine is a high-grade heat resource. With temperatures exceeding 600°C, the exhaust gases are at higher temperatures than is used in steam-cycle plants. It makes sense to use the high-grade "waste heat" of an open-cycle gas turbine, and this is exactly what is done in a "combined cycle" gas turbine plant. Here, hot exhaust gases produced by a gas turbine are passed through a boiler, producing high-pressure steam which is then expanded to produce additional power in a steam turbine.

Combined cycle power plants contain a gas turbine and a small steam-cycle power generator, and convert about 60% of the fuel energy into electrical power output. As you might expect, combined cycle generating plants are more complex and more costly to build than an open-cycle gas turbine plants, but consume much less fuel (and produce much less emissions) to generate the same electrical power output.

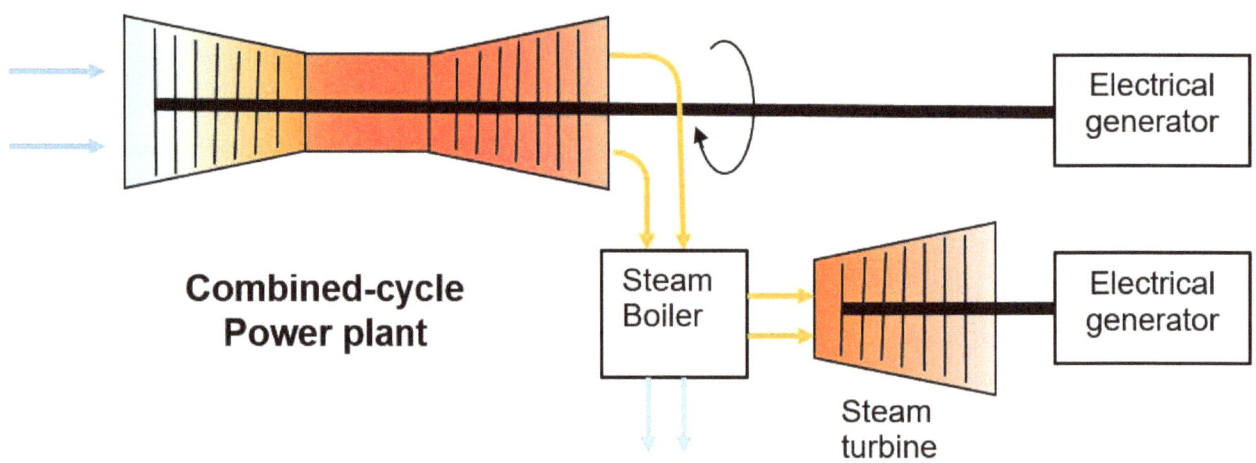

In the past, nearly all power plants using natural gas were open-cycle gas turbines, as these were cheap to build (but inefficient and costly to operate). However, in the last ten years, combined cycle plants have become common.

Low-temperature geothermal power generation

In striving to achieve greater fuel efficiency for electricity generation, engineers design power stations to operate with the highest temperatures allowed by the materials and technology available. However, in some cases, the maximum temperature is determined by the available heat source, which may be low-grade "waste heat". An interesting example is a low-temperature geothermal generating station in the remote town of Birdsville in outback Queensland. The town relies for its water supply on bore water from an artesian aquifer deep underground. The water reaches the surface at nearly 100°C, just below the normal boiling point of water. The hot bore water needs to be cooled before it can be circulated to town residents, so it comprises a free source of unwanted "waste heat". At the same time, electricity for the town was produced by engines burning diesel fuel, a very expensive energy source.

A low-temperature Rankine cycle engine was installed – basically a refrigerator operated in reverse, exactly as depicted in the simple diagram at the beginning of this chapter. Hot bore water was passed through a boiler, where liquid propane boiled into vapour. The propane vapour was expanded to produce power, condensed into liquid at air temperature, and then recycled through a pump to the boiler.

In this case, the temperature of the heat source was limited to the 100°C temperature of the bore water, and the heat sink was the local environment at ambient temperature. The air temperature in Birdsville is generally above 30°C in daytime, and often exceeds 40°C in summer. Consequently, the engine had to operate within a narrow temperature range between 40°C (313° above absolute zero) and 100°C (373° absolute). The maximum possible efficiency that could be achieved for such a power station is (373°-313°)/373° = 16%.

In fact, the actual efficiency would be significantly less than 16%. The power station contained a single heat engine to minimise complexity and cost, so its boiler extracts heat from the hot bore water at a single temperature. What would be the best temperature to operate the boiler? Let's consider two options:

- We could extract heat from the hot water at, say, 98° – just below the temperature of the bore water. This would allow the highest possible efficiency of the heat engine, but we would only extract a tiny fraction of the heat energy in the water as it cooled from 100° to 98°.
- Alternatively, we could operate the boiler at, say, 42°C – just above ambient air temperature. This would allow the boiler to extract nearly all of the heat energy in the hot water, but the efficiency of the engine would be extremely low.

It should be evident that maximum power would be produced by extracting heat at some intermediate temperature roughly halfway between 100°C (the available T_{HOT}) and 40°C (T_{COLD}). For readers who completed first-year university calculus, it is a nice exercise[See Note 1] to derive that the optimal temperature that allows maximum power production is $\sqrt{T_{HOT} T_{COLD}}$ = 68°C. At this boiler temperature, a maximum of 6.7% of the heat content of the bore water can be converted into electrical power output.

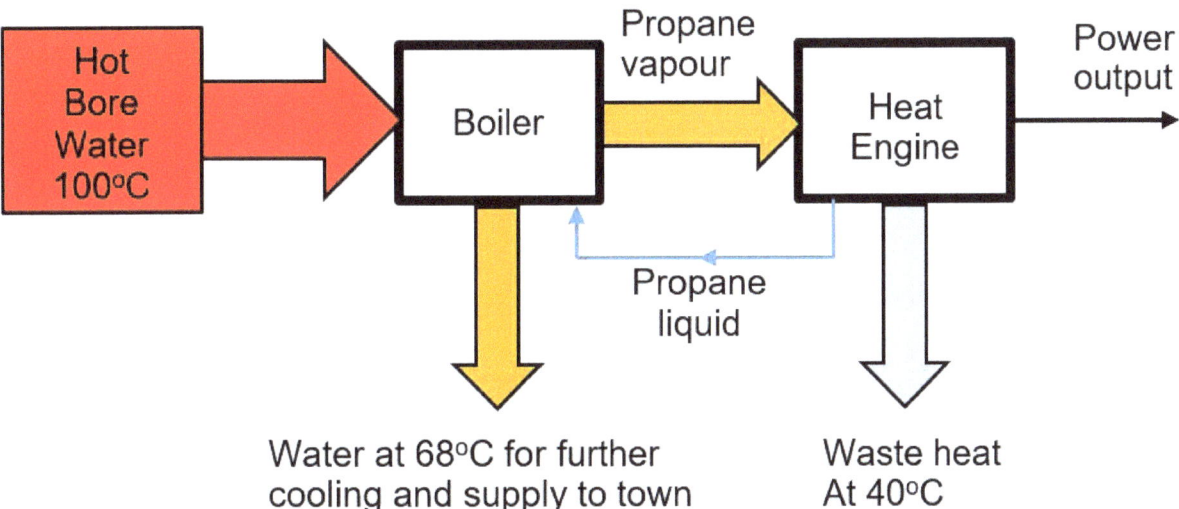

Configuration for extracting heat from hot bore water using a single heat engine. The width of the arrows indicates the relative energy content in the stream.

The actual efficiency of the power plant would be even less than this, but since the heat source is literally available at zero cost (and would otherwise be discarded), and the alternative diesel generation is very expensive, low-temperature geothermal power generation proved cost-effective and operated for many years in Birdsville.

Solar power generation

Throughout the 4.5 billion year existence of the Earth, sunlight has been the major energy source for the planet and its lifeforms. Solar energy was used by ancient plants and algae living in shallow seas to produce carbohydrates, proteins and oils. Some of this biomass was converted into coal, oil and gas when deposits accumulated and were buried in sediments as the seafloor subsided over millions of years.

Sunlight creates temperature differences across the Earth's surface, and the atmosphere acts as a heat engine to redistribute heat from equatorial regions towards the poles. This gives rise to winds, waves, rains and storms.

In recent years, solar energy has been tapped directly as an energy source for humanity. Two types of technologies are used to generate electricity from sunlight:

- Solar thermal power generators concentrate sunlight onto a collector, with the collected heat converted to mechanical power (and then, electricity) in a heat engine.
- Photovoltaic power generators convert solar radiation directly into electrical energy.

We can think of the entire solar system as a huge heat engine with the sun as a heat source. The surface of the sun has a temperature of about 6,000° (absolute). Each square metre at the sun's surface emits about 70 million watts of visible, ultraviolet and infrared radiation. As the radiation moves outwards and spreads across the solar system, its intensity reduces with the inverse square of the distance. As it travels from the sun's surface to the orbit of the Earth (about 200 times further from the centre of the sun), the light spreads across an area some 40,000 times larger. By the time solar radiation reaches Earth, its intensity has fallen to about 1,300 watts/m^2. Some is scattered by the Earth's atmosphere, and most ultraviolet radiation is absorbed, so that sunlight striking perpendicular to the Earth's surface has an intensity of about 1,000 watts/m^2.

It is possible, using mirrors or lenses, to concentrate sunlight to increase its intensity. High temperatures are produced when concentrated solar radiation is absorbed by a collector plate. In theory, it would be possible to produce temperatures as high as 6,000° – as hot as the temperature at the surface of the sun. It is not possible to produce even higher temperatures by cleverly focussing and concentrating solar radiation, as this would transfer heat from the surface of the sun to an even hotter collector on the Earth – and this would violate the Second Law of Thermodynamics.

Mind you, with current technology, it would not be possible to build a solar collector that could withstand 6,000°. All common materials vaporise, burn or decompose at temperatures well below this. Nonetheless, it is instructive to consider what would happen if new materials could be developed in the future that could withstand such temperatures.

As we focussed highly concentrated solar radiation on our hypothetical absorber, its temperature would rise and it would begin to emit "blackbody radiation" from its surface - just as an object heated in a furnace begins to glow red hot. The intensity of emitted radiation rises rapidly with increasing temperature. With each 1% increase in absolute temperature, the intensity of emitted radiation increases by 4%. As the temperature of the absorber approaches 6,000°, the intensity of emitted radiation becomes nearly as intense as the highly concentrated sunlight. If the "absorber" reached 6,000°, it would emit radiation (and lose energy) at the same rate as it absorbed concentrated sunlight. At even higher temperature, the "absorber" would emit more radiation than it absorbed, and its temperature would fall.

Let's consider the maximum possible efficiency of a solar thermal power station, which is basically a heat engine using a solar collector as a heat source. The maximum temperature of the heat source is 6,000° (the temperature at the surface of the sun), and the heat sink is the surrounding environment on Earth (at a temperature of around 300° absolute). Consequently, it would ultimately be possible for **absorbed solar radiation** to be converted into useful work with an efficiency of (6,000-300)/6,000 = 95%. But this does not mean that 95% of **incident solar radiation** can be converted into work. A solar thermal power station faces the same type of dilemma as we encountered with a low-temperature geothermal power station. The question to consider is this: What is the optimal temperature at which solar radiation should be absorbed?

To illustrate the problem, let's again consider two hypothetical options:

- We could operate the solar collector at just below the maximum possible temperature of 6,000°. In principle, an ideal heat engine could convert this heat into work with 95% efficiency. But, at this temperature, the collector would be emitting radiation at nearly the same rate as the concentrated solar radiation was being absorbed. Only a small fraction of the incident solar energy would be available as a heat source.
- On the other hand, we could operate the collector at low temperature – just above ambient temperature. Then, radiation emitted by the collector would be negligible, and nearly all incident solar energy would be available to operate the heat pump. But, at this low temperature, the collected heat could only be converted at low efficiency into useful work.

Once again, It should be evident that maximum power would be produced by operating the solar collector at some intermediate temperature, well below the theoretical maximum of 6,000° and well above ambient temperature. As it turns out that a maximum of 86% of incident solar energy could (in theory) be converted into electricity, and this would be achieved at a collector temperature of around 2,500° (absolute).

This limit has been derived for most efficient solar thermal power station that is allowed by the 2nd Law of Thermodynamics), but the same efficiency limit also applies to photovoltaic power generators, even though they do not appear to be "heat engines" in the usual sense.

We can again conduct a "thought experiment" to show that a photovoltaic solar module cannot be more efficient than the maximum possible efficiency for a solar thermal power station.

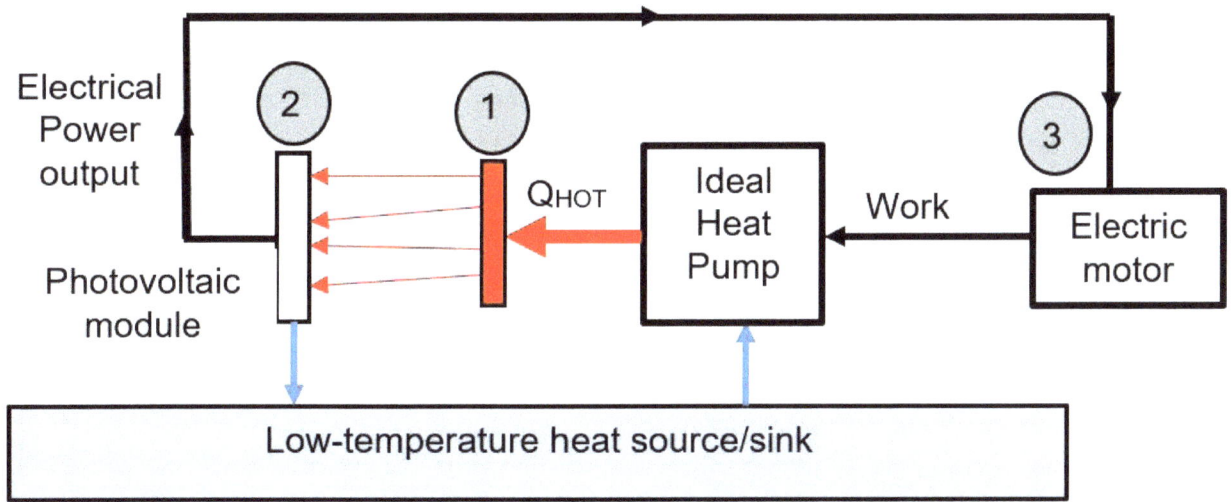

Imagine the hypothetical scenario shown here, in which we have a solar thermal power station that is as efficient as allowed by the 2nd Law of Thermodynamics. Such an ideal heat engine could – in principle - operate reversibly as a heat pump, and could produce high temperatures (simulating conditions at the surface of the sun), causing radiation to be emitted (step 1). Radiation emitted at this temperature is focussed onto a photovoltaic module, which generates electricity (step 2). The electrical power output of the photovoltaic module is used to drive the heat pump (step 3), so that heat and electrical energy are continually cycled through the system.

If the photovoltaic module were more efficient than our ideal heat pump, electricity produced by the photovoltaic module would be greater than the energy required to operate the heat pump. As energy was cycled through the system, more and more work output could be produced, without any additional energy input. This is a nonsensical result, and could not happen because it would violate the 2nd Law of Thermodynamics. We can thus conclude that photovoltaic modules are subject to the same 86% limit on the efficiency of converting solar energy to electricity.

In fact, with current technology, the actual efficiency of solar thermal power stations and photovoltaic modules is about one-quarter of the maximum possible efficiency, so there is still considerable scope for technological improvement.

Note (1): Calculation of optimal efficiency

We want to produce maximum power from bore water containing heat energy Q_{HOT}. Let's say that the boiler extracts heat from the bore water at temperature T, which is intermediate between T_{HOT} and T_{COLD}. Since the heat capacity of water (the amount of heat released as water cools by one degree) is relatively constant, the amount of heat recovered from the bore water is then $Q_{HOT}[\frac{T_{HOT}-T}{T_{HOT}-T_{COLD}}]$. Of this heat energy, the fraction that can be converted into useful work is limited to $[\frac{T-T_{COLD}}{T}]$, so the maximum work output is $Q_{HOT}[\frac{T_{HOT}-T}{T_{HOT}-T_{COLD}}][\frac{T-T_{COLD}}{T}]$.

We can find the value of temperature T that yields the highest work output by taking the first derivative of the power output and setting this equal to zero. Solving for the optimal value of T gives $\sqrt{T_{HOT} T_{COLD}}$. If the temperature difference $T_{HOT} - T_{COLD}$ is small in relation to the absolute temperature T_{HOT}, then it turns out that the optimal temperature is halfway between T_{COLD} and T_{HOT}. Substituting this optimal temperature for the conditions T_{HOT} = 100°C and T_{COLD} = 40°C gives a maximum power output equal to 6.7% of the heat energy in the stream of hot water.

24. Internal combustion engines

In the 18th and 19th centuries, wood and coal were the dominant fuels powering the industrial revolution. These fuels were burned in steam engines to produce mechanical power to pump water, grind grain, move goods by rail, and many other applications. Steam engines are one type of **external combustion engine**, in which the fuel is burned in a boiler to produce heat, and then the heat is transferred to a "Working fluid" (such as water/steam) inside a separate heat engine.

Early steam engines were grossly inefficient, consuming large amounts of wood or coal to produce a modest power output. These engines used vacuum created by condensation of steam – just like the imploding can demonstration described in a previous chapter. The crude technology available was suited for very large engines operating at very low speed, producing relatively small amounts of power. These engines could produce large forces by atmospheric pressure acting on large pistons, but they converted less than 1% of fuel energy to mechanical work output.

Design improvements and improved materials and technology allowed the efficiency to be improved several-fold. After about 1800, steam engines used expansion of high-pressure steam, rather than condensation of steam, which was inherently more efficient.

The technology and efficiency of steam engines improved dramatically through the 19th century, and it became practical to use steam engines in a wide range of new applications. Ironically, development of more efficient steam engines led to greatly increased consumption of coal. However, it was gradually becoming clear that further increases in efficiency were not simply limited by the technology available, but also by fundamental physical laws (the 2nd Law of Thermodynamics). As we've seen, the maximum possible efficiency of any heat engine is limited by the temperature of the heat source T_{HOT} and the heat sink $T_{AMBIENT}$. In 1824, the French physicist and military engineer Nicolas Sadi Carnot first derived that the maximum efficiency was given by $(T_{HOT}-T_{AMBIENT})/T_{HOT}$. This is often referred to as the "Carnot efficiency limit".

Starting around 1850, petroleum (oil and gas) started to be produced in large quantities, and these became the dominant fuels of the 20th century and beyond. Liquid fuels (such as gasoline, diesel and kerosene) can readily be stored and pumped, and are extremely convenient to use (you don't need someone to shovel the fuel into an engine). In the late 19th century, **internal combustion engines** were developed that could utilise these fuels. In these engines, the fuel is mixed with air, and the air-fuel mixture serves as the "working fluid" of the engine as well as providing heat when it burns.

The development of the gasoline engine used in most of our cars can be traced back directly to the engine developed by the German inventor Nikolous Otto around 1886. It comprises cylinders containing pistons which slide to draw in fuel-air mixture, compress the gas, burn the compressed fuel-air mixture, (further increasing its temperature and pressure), expand the gas, and then exhaust the gas to the atmosphere. Each piston slides back-and-forth within its cylinder (with a gas-tight seal between the sides of the piston and cylinder) as the crankshaft of the engine rotates.

For each two rotations of the engine, each cylinder/piston combination undergoes 4 strokes in succession.

1. **Intake stroke**

    With the intake valve open, the piston moves downwards within the cylinder, drawing in fresh air-fuel mixture at ambient temperature (T_1).

2. **Compression Stroke**

    With all valves closed, the piston moves upwards, compressing the gas. Typically, the gas is compressed to about one-tenth of its initial volume. This increases the pressure and temperature of the gas (to T_2).

3. **Expansion (Power) Stroke**

    A spark ignites the air-fuel mixture, which rapidly burns, greatly increasing the temperature of the gas (to T_3) and increasing its pressure. At the high temperatures and pressures, the air-fuel mixture burns almost instantaneously, causing the temperature and pressure to rise dramatically (about 2-3 times). The piston is driven downwards by the high-pressure gas, which expands and cools to temperature T_4. Note that expansion ratio of the gas (the ratio of final to initial volume) is exactly the same as the compression ratio in the previous compression stroke.

4. **Exhaust Stroke**

    With the exhaust valve open, the piston moves upwards, pushing the combustion gases out of the cylinder (in preparation for the next cycle).

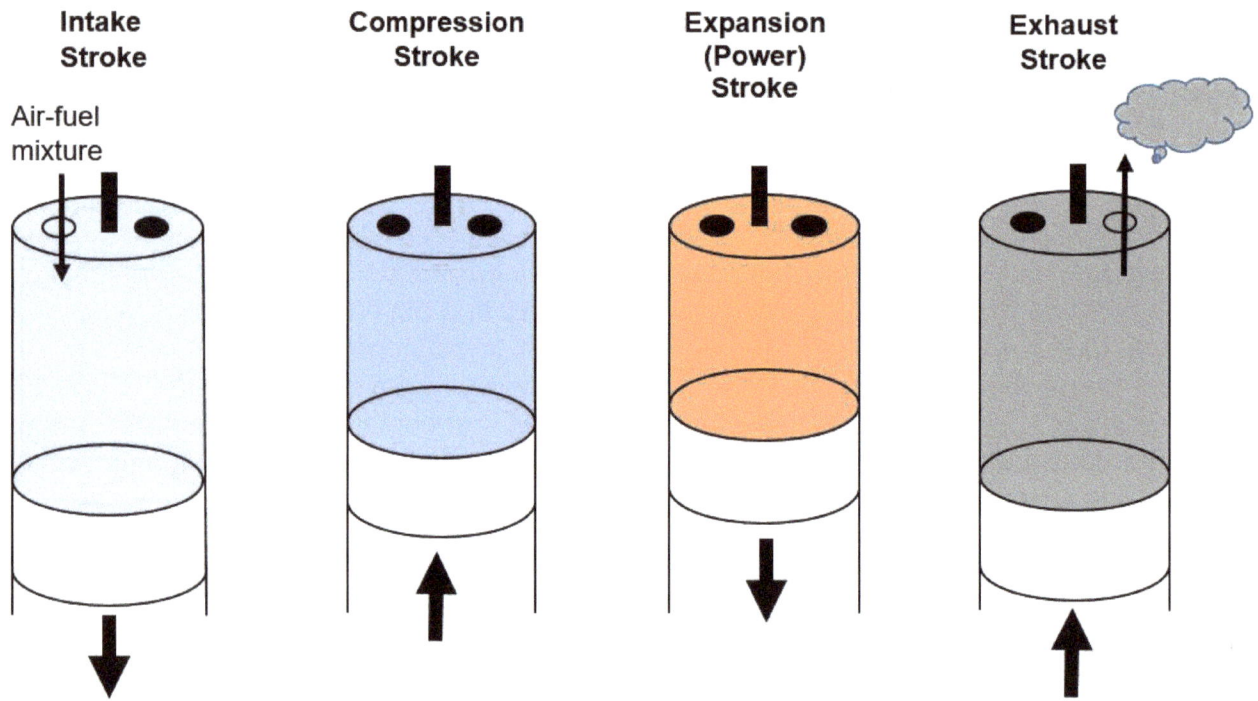

This basic 4-stroke cycle is termed the Otto cycle. It is illustrated very well in the following short animated videos:
www.youtube.com/watch?v=2Yx32F1cncg
www.youtube.com/watch?v=OGj8OneMjek

To understand the operation of an Otto cycle engine, it is useful to draw a graph showing how the pressure and volume of the air-fuel mixture change during each stroke of the cycle. A pressure-versus-volume graph is shown below. Note that the work required to compress the gas (during stroke 2) is given by the area under the pressure-volume line depicting the compression stroke. The work produced during the Expansion/Power stroke is given by area under the pressure-volume line depicting the expansion stroke. The net work done by the gas during the cycle (Work done by expanding gas – Work required to compress the gas) is given by the shaded area between the two lines.

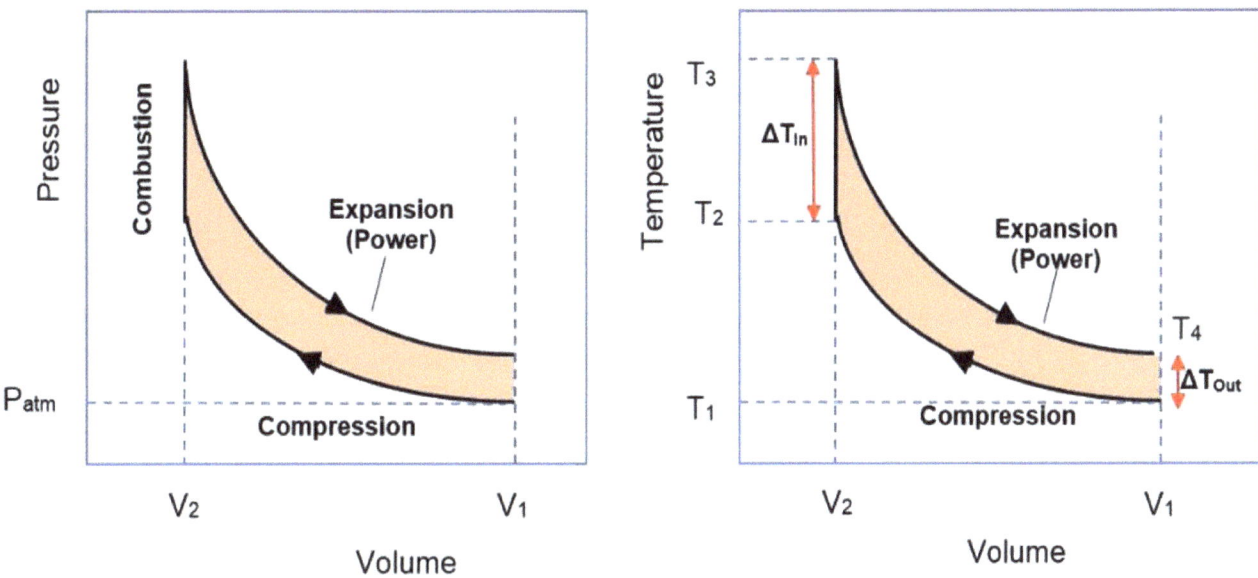

The extremely hot combustion gases lose heat when they contact the relatively cool walls of the cylinders, and this heat loss accounts for a major inefficiency of internal combustion engines. Typically, as much as one-third of the heat released by combustion of the fuel is lost through the wall of the cylinders, absorbed by the coolant and dissipated to the atmosphere. Additional losses are caused by friction as the pistons slide within the cylinders, and in other moving parts. These losses can be minimised by careful design of the engine and choice of materials and lubricants.

However, other losses are inherent in the operation of internal combustion engines, and these losses cannot be minimised or eliminated directly by better engine design and materials. These losses result from a basic feature of Otto-cycle engines – the compression ratio of the air-fuel mixture (the ratio of the initial to the final volume of the gas before compression) is **exactly the same** as the expansion ratio for the combustion gases.

To illustrate, let's assume that the air-fuel mixture is compressed to one-tenth of its initial volume. We can readily calculate[see Note 1] that this compression increases the absolute temperature of the gas by 2-1/2 times (from, say, 300° to 750°). Combustion of the fuel will further increase its temperature to, say, 1,500°. The hot combustion gases are then expanded to produce mechanical power, converting its heat energy to useful work. Typically, one-third of the heat energy released by the burning fuel remains in the hot exhaust gases which are discarded to the atmosphere.

As it turns out [also Note 1], when a gas is compressed or expanded, the ratio of its final/initial temperature is determined by the ratio of its final/initial volume. But since **the compression ratio is the same as the expansion ratio**, then the ratio of gas temperatures during the compression stroke is the same as the ratio of gas temperatures during the expansion stroke. Referring to the graph of temperatures during the cycle, note that:

$$\frac{T_2}{T_1} = \frac{T_3}{T_4}$$

The temperature ratio T_2/T_1 during the compression stroke is exactly equal to the temperature ratio T_3/T_4 during the expansion stroke.

So let's see what happens in an engine with a 10:1 compression ratio. Let's say that the outside air temperature T_1 is 27°C (300°Kelvin). Compression of the air-fuel mixture to one-tenth its volume heats it to T_2 = 750°K (so, T_2/T_1 = 2.5). Let's say that combustion of the air-fuel mixture heats the gas by an additional 1,000°, from 750°K to 1,750°K (T_3). Since temperature ratio T_3/T_4 must also equal 2.5, the exhaust gas temperature T_4 will be 1,750/2.5 = 700°K. The amount of heat energy retained in the exhaust gas is proportional to the 400 degree temperature difference between the exhaust gas temperature (700°K) and the ambient air temperature (300°K). Compare this with the amount of heat energy put into the gas during combustion, which is proportional to the 1,000° rise in temperature. The exhaust gas contains 40% of the heat energy that was released during combustion of the fuel, and the maximum possible efficiency of the engine is 60%.

Applying the same arguments in general, it is not hard to derive that the maximum possible efficiency of an Otto cycle engine is:

Equation (1A) Maximum possible efficiency = $\dfrac{T_2 - T_1}{T_2}$

This may also be written as:

Equation (1B) Maximum possible efficiency = $1 - \dfrac{T_1}{T_2}$

Where T_1 refers to the outside air temperature, and T_2 is the temperature of the air-fuel mixture immediately after compression.

But note that the ratio T_1/T_2 is determined only by the compression ratio of the engine (according to the equation given in the following Note 1). The higher the compression ratio, the greater is the maximum possible efficiency of the engine.

For maximum possible efficiency, we would like the compression ratio to be as large as possible. However, there is a limit to how high the compression ratio can be for "spark ignition" engines. If the compression ratio is too high, the air-fuel mixture detonates, producing shock waves rather than a smooth increase in pressure. The vibration produced by such detonation, called "knocking", can damage the engine. Also, if the compression ratio is too high, the air-fuel mixture gets so hot during the compression stroke that the fuel-air mixture ignites and burns before the cylinder reaches the top of the cylinder. "Pre-ignition" occurs.

The diesel engine offers one way to get around this limitation. In diesel engines, fuel is not injected into the cylinder until the air has been compressed and the piston has reached the top of its stroke. In fact, diesel engines rely on the very high temperatures produced by compression to ignite the fuel. Diesel engines use very high compression ratios (about 18:1) to ensure that the air gets so hot during compression that the fuel instantly ignites, without a spark, when it is sprayed into the cylinder at the end of the compression stroke.

A diesel engine is referred to as a "compression ignition engine". "Knocking" does not occur because combustion takes place at the surface of fine droplets of fuel. Pre-ignition cannot occur in a diesel engine because fuel is not injected until the piston reaches the stop of the compression stroke. Diesel engines operate at much higher pressures than spark ignition (petrol) engines, and thus are manufactured with thicker cylinder walls and stronger components. They tend to operate at slower rotational speeds, and are heavier (as well as more expensive) for the same power output.

Other factors besides compression ratio impact on the efficiency of internal combustion engines. Engines are most efficient when operated at steady conditions at their rated power output. However, in most real-world applications, engine speeds and loads are constantly varying, and actual engine efficiencies are often much less than the maximum efficiency at rated power output. In particular, the speed and load on a car engine varies markedly as a car drives in city traffic, and this significantly impacts on its fuel efficiency. Most cars use significantly more fuel (per kilometre travelled) for city driving than to maintain constant speed on the highway, even though air resistance is reduced at lower speeds.

A general rule of thumb for an engine operated continuously at rated load is that:
- One-third of the fuel energy is converted to useful work output (a bit less for petrol engines, a bit more for diesel engines).
- One-third of the fuel energy is retained as heat in the hot exhaust gases.
- One-third of the fuel energy is lost as heat to the cooling system and dissipated in the radiator.

Another type of internal combustion engine is the jet engine, or gas turbine. Gas turbines use entirely different technology to achieve a very similar thermodynamic cycle, with an identical result for the maximum possible efficiency. Instead of using pistons sliding within cylinders, a gas turbine uses a series of rapidly rotating turbine blades to compress the gas. The compressed gas flows into a combustion chamber, where fuel is injected and burned, increasing the temperature of the gas. The very hot gas is then expanded in another set of rotating turbine blades.

The technology, construction and materials used in gas turbines are completely different from Otto-cycle and diesel engines, and they look completely different. But as far as the basic operating cycles of these engines, they are pretty similar. In all cases, air (the "working fluid") is taken into the engine, compressed,

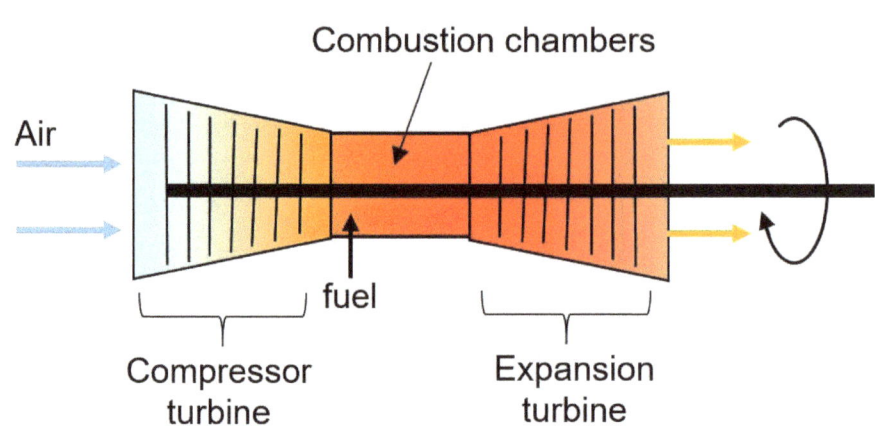

heated by combustion of a fuel, and then the hot gas is expanded to produce mechanical work. The power output of the engine is the difference between the work required to compress the gas and the work produced during expansion of the gas.

From a thermodynamic perspective, the big difference is that:

- In an Otto-cycle or diesel engine, combustion occurs when the air-fuel mixture is confined within a small volume when the piston is at the top of its stroke. **Combustion occurs at constant volume**, and **the ratio of initial/final gas volumes is the same** during compression and expansion of the working fluid.
- In a gas turbine, combustion gases leave the combustion chamber at the same pressure as air and fuel enter. **Combustion occurs at constant pressure**, and **the ratio of initial/final gas pressure is the same** during compression and expansion of the working fluid.

The cycle for a gas turbine (or jet engine) is called the Brayton cycle.

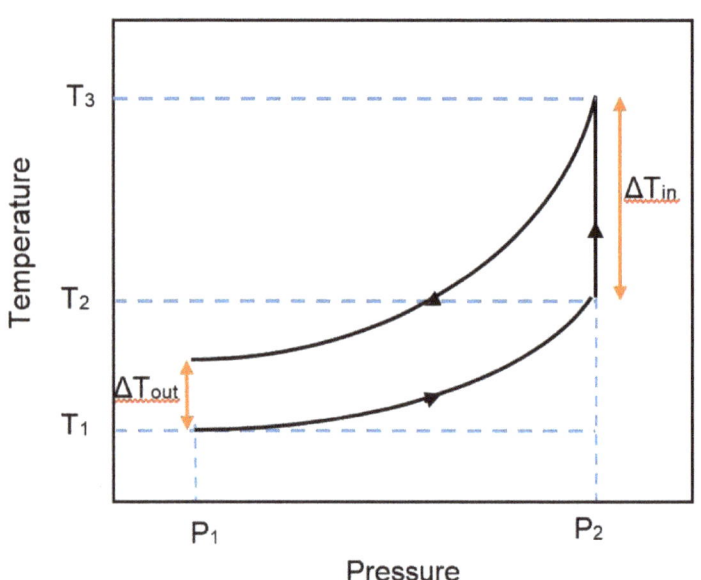

From the perspective of achieving the maximum possible efficiency, a gas turbine engine is subject to the same limitation as applies to an Otto or diesel cycle engine, as given by Equation (1). The efficiency is determined by the ratio of gas temperatures before and after compression, but in this case, the temperature ratio T_2/T_1 depends upon the **ratio of gas pressures** before and after compression. The main barrier to achieving higher efficiencies through higher pressure ratios is the ability of the turbine blades to withstand the extremely high temperatures of the exhaust gases.

Generally, gas turbines tend to be slightly less efficient than diesel engines, but they offer the advantage of being extremely light and compact. Gas turbines are well suited to applications requiring very high power/weight ratios and very compact power sources. Thus, gas turbines are widely used in aircraft, military vehicles (such as the M1A1 tank, the standard tank of the US Army), and some locomotives and high-speed ships.

The maximum possible efficiency for an Otto Cycle or a Brayton Cycle engine is given by Equation (1), which *looks* like the equation giving the maximum possible efficiency for any heat engine allowed by the Second Law of Thermodynamics (the Carnot efficiency limit discussed earlier). But looks are deceiving! The maximum efficiency of an Otto or Brayton cycle engine is *less* than the theoretical maximum for a heat engine. The efficiency is limited by the temperature of the air-fuel mixture after compression (T_2), but the temperature of the "heat source" (the burning fuel) is well above this temperature (perhaps a thousand degrees higher!). If we look at the engine from the other side – the heat sink, an ideal heat engine will get rid of its waste heat at lowest possible temperature (the temperature of the surrounding air). For an Otto or Brayton cycle engine, waste heat in the exhaust gases has temperatures well above the outside air temperature (typically, about 500°C for a fully-loaded petrol or diesel engine, or about 700°C for a gas turbine).

It is possible to improve the efficiency of an Otto or Brayton Cycle Engine, more closely approaching the maximum possible efficiency of a heat engine, by utilising the high-temperature waste heat carried in the exhaust gases. This is now done in Combined Cycle Gas Turbines, which pass the hot exhaust gases of a gas turbine through a boiler to produce high-pressure steam, which is then expanded through a steam turbine. Gas turbines are particularly well suited to combined cycle plants, since all of the waste heat is contained in the exhaust gases, which have very high temperatures (600°C or more). Nonetheless, it is also possible to utilise the waste heat in the exhaust gas of Otto-cycle or diesel engines, although this is rarely done in practice. Because combined cycle plants contain an internal combustion engine and a separate heat engine, they are more complex and more expensive than conventional "open cycle" plants. However, the higher initial cost is offset by lower fuel consumption and reduced fuel costs.

Note 1: Temperature ratios for expansion or compression of a gas

Whenever a gas is compressed or expanded, the initial and final temperature of the gas is related to its initial and final volume. The following equation can be calculated by assuming that the gas behaves as an "ideal gas" (as indeed most gases do, except at extremely high pressure or very low temperature), and assuming that no heat is exchanged with the surroundings during the expansion or compression. The mechanical work involved in compressing (or expanding) the gas is equated with its gain (or loss) of heat energy, and then integrating the equation from the initial volume V_1 to its final volume V_2, and from the initial temperature T_1 to its final temperature T_2. The result is:

$$\frac{T_2}{T_1} = \left(\frac{V_1}{V_2}\right)^{R/Cv}$$

    Where **T_1** and **T_2** are the initial and final temperatures
        V_1 and **V_2** are the initial and final volumes
        R is the Universal Gas Constant (8.3 Joules/mole-degree)
        Cv is the heat capacity of the air mixture (approx. 2.5 R, or 20 Joules/mole-degree)

25. Feeding the world's population

The climate, atmosphere and other conditions on the surface of Earth are ideally suited for life as we know it. Perhaps that should be no surprise, as life on Earth has evolved to suit the particular environments on Earth. To some extent, plant and animal species alter their environment to suit their own needs (and mankind has certainly done this to a huge extent, and not always to the benefit of other species, the environment and future generations).

In particular, all life on Earth with which we are familiar derives energy from sunlight striking the Earth's surface (some microbes derive energy from geochemical sources at deep sea vents or within fissures of rocks deep underground, but these biological communities are isolated from the world we know and are largely unseen and little known).

We humans obtain energy from the food that we eat, largely from carbohydrates (sugars and starches) produced by plant crops. We also employ grazing animals to digest cellulose (and other carbohydrates that we cannot digest) obtained from grass and straw to produce meat and milk. Our food sources also include fish which consume aquatic plants, or eat other fish which consume aquatic plants. So, human civilisation and life as we know it is based entirely on plants harnessing solar energy in a process called "photosynthesis".

In man-made solar power stations, the energy of the sun produces mechanical power or electricity, but in photosynthetic plants, solar energy is used to drive chemical reactions. Plants react carbon dioxide from the air with water to form carbohydrates. The reaction also produces a by-product of oxygen, which is released to the atmosphere.

$$\text{Photosynthesis:} \quad \underset{\text{Carbon dioxide}}{CO_2} + \underset{\text{water}}{H_2O} \rightarrow \underset{\text{Carbohydrate}}{CH_2O} + \underset{\text{Oxygen}}{O_2}$$

Energy is required for the photosynthesis reaction to occur. An energy input of 16 megajoules must be provided for each kilogram of carbohydrate produced.

Humans, like all animals, derive energy by reacting carbohydrates with oxygen, the reverse of the photosynthesis reaction. This reaction, called "respiration", occurs in the cells of our bodies. Carbohydrate (absorbed from our food) is reacted with oxygen (absorbed from the air by our lungs) to produce carbon dioxide and water. The reaction releases 16 megajoules of energy per kilogram of carbohydrate consumed. We use this energy to move, to build or repair damaged tissue, to pump blood, breath and to operate our brains.

$$\text{Respiration:} \quad CH_2O + O_2 \rightarrow CO_2 + H_2O$$

Nearly all of the energy used by humans, as individuals and as an industrial society, is ultimately derived from solar energy that was captured by plants at some time in the past.

Like solar photovoltaic modules, photosynthetic plants are "heat engines", although the biochemical processes used by plants are completely different from man-made engines. Plants derive energy from solar radiation emitted from the surface of the sun (at 6,000°), and dissipate waste heat to the surrounding air (at about 300° absolute). Consequently, plants are restricted to the same 86% maximum conversion of incident solar energy that applies to solar power stations. But the actual solar conversion efficiency of real plants is nowhere near that high. Most plants convert only 1-2% of the energy in incident sunlight into chemical energy of sugars, starches and other carbohydrates.

You may be surprised to find, as I was, that after billions of years of evolution, plants did not evolve to utilise sunlight very efficiently. The simplest explanation is that plants **don't need** to utilise sunlight efficiently. In most climate regions (especially throughout Australia, for example), the intensity and duration of sunlight is far more than plants need, or can use. Generally, the growth of land-based plants is not limited by the availability of sunlight, but by the availability of carbon dioxide and water (See Note 1 for an exception). And, as we'll see, the ability of terrestrial plants to absorb carbon dioxide from the air is limited by the availability of water. I used to think that plants were biochemical machines that convert sunlight into carbohydrate, but this gives a misleading impression of the key challenge faced by plants and what they need to do to survive and grow.

Both plants and animals derive key requirements for life from the air. Animals breathe air into their lungs to absorb oxygen (and to dispose of carbon dioxide). Plants use the huge surface areas of pores within their leaves to absorb carbon dioxide (and exude oxygen). However, **plants have a much more difficult job than animals**. Oxygen needed by animals comprises 21% of the air, but carbon dioxide needed by plants is only 0.04%. The concentration of carbon dioxide is **500 times less** than the concentration of oxygen in the Earth's atmosphere, so the job of plants is 500 times harder! Plants must access huge volumes of air to get the carbon dioxide they need to grow. To meet this challenge, plants produce an extensive canopy of leaves containing pores with an enormous surface area. Microscopic openings in the underside of the leavers, called "stomata", allow carbon dioxide to diffuse into air spaces inside the leaf.

Within the porous structure of a leaf, gases are exchanged across membranes which – as in all living organisms – are composed of cells which consist largely of water. Water vapour can diffuse outwards through the membranes, as carbon dioxide can diffuse inwards. This poses a fundamental dilemma for plants.

Let's consider what happens within the pores of a leaf. We'll assume that the leaf and surrounding air are at a temperature of 25°C, at which the equilibrium vapour pressure of water is 3,200 Pascals (about 3% of atmospheric pressure). Gases inside the pores becomes saturated with water vapour at this temperature, so the "partial pressure" of water vapour inside a pore is 3,200 Pascals. This, by the way, is eighty times greater than the partial pressure of

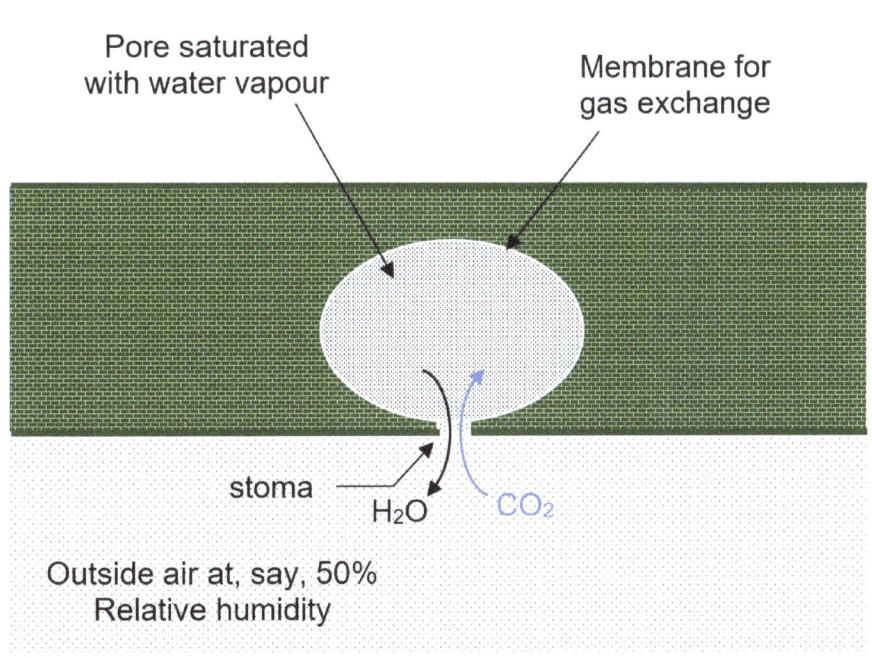

carbon dioxide in the air outside. This means that there are eighty times more water vapour molecules per unit volume within pores of the leaf than there are carbon dioxide molecules in the air outside. Herein lies the conundrum for the plant.

Of course, air outside the leaf also contains some water vapour, and the rate of escape of water molecules depends on the relative humidity. At a typical relative humidity of 50%, the partial pressure of water vapour in the air is half the equilibrium vapour pressure (or 1,600 Pascals). The **net rate** that water molecules diffuse out through the stoma opening is proportional to the **difference** in water vapour pressure, which is 1,600 Pascals.

At the same time, since the pressure of carbon dioxide in the outside air is 0.04% of atmospheric pressure, the maximum difference in the partial pressure of carbon dioxide across the stoma openings is 40 Pascals.

The forty-fold greater difference in water vapour concentration would cause water vapour to evaporate and be lost from plant tissues forty times faster than carbon dioxide can be absorbed. In fact, the rate of water loss will likely be even greater. Water molecules have only 41% of the mass of carbon dioxide molecules (Note 2), and thus, water molecules move at greater velocity and diffuse about 50% faster than carbon dioxide molecules. Taking this factor into account, the plant loses at least 60 molecules of water for every molecule of carbon dioxide absorbed from the air.

Consequently, at an air temperature of 25°C and 50% relative humidity, a plant must lose at least 40 litres of water (through evaporation, or "transpiration") for each kilogram of carbohydrate produced. The minimum water requirement rises rapidly with increasing air temperature (reflecting the exponential variation of water vapour pressure with temperature), and with the dryness of the air.

Let's see how this calculated figure compares with the water consumption of real plants. The most water-efficient plants are cacti that have evolved in deserts and arid regions of the world. These plants have developed complex chemical mechanisms to minimise water use. In particular, these plants have developed separate enzyme systems to absorb carbon dioxide at night, when temperatures are cooler, and then produce carbohydrate during daytime, when sunlight provides the energy needed for photosynthesis. Plants with this "CAM metabolic pathway" require about 50 litres of water to produce each kilogram of carbohydrate. This is in very close agreement with the calculated minimum water requirement, based on typical night-time temperatures of 25° and humidity of 30-50%.

But desert plants using the CAM metabolic pathway pay a price for their frugal use of water. When adequate moisture is available, CAM plants are not able to grow as rapidly or compete with plants that have evolved to take advantage of wet conditions. These other plants, which are far more water-intensive, are the ones we depend on for food and other agricultural products.

More common is the "C4 metabolic pathway", used by many grasses, maize, sugarcane and sorghum. These plants absorb carbon dioxide during the day, when sunlight provides the energy to produce carbohydrate, and apparently lack other water-saving strategies employed by arid-region plants using the CAM metabolic pathway. C4 plants generally require about 300 litres of water to produce each kilogram of dry biomass.

Most trees, wheat, soybean, alfalfa, rice and many other crops use a "C3 metabolic pathway". This pathway is the simplest mechanism for photosynthesis, and does not employ a separate enzyme system to absorb carbon dioxide at low concentrations. C3 plants typically require about 600 litres of water to produce each kilogram of biomass. Additional water losses occur due to evaporation from the soil and non-optimal application of water to the root zone of the plants.

Depending on agricultural practice and climate, water consumption is often within the range of 1,000-2,000 litres for each kilogram of carbohydrate produced (dry basis). Typically, when you buy a one kilogram bag of dried rice, it "embodies" the consumption of about a tonne of water in its production.

Some commentators argue that the water efficiency of plants will be improved by rising levels of carbon dioxide in the atmosphere due to extensive burning of fossil fuels. The concentration of carbon dioxide in the Earth's atmosphere has already increased 40% since the onset of the industrial revolution, and continues to rise. And indeed, higher levels of carbon dioxide is found to generally improve the water efficiency of plants. However, this will likely be more than offset by higher global average temperatures, increased threat of crop damage by severe storms and droughts, and increased losses due to insect pests and weed species. Consequently, global climate change arising from rising carbon dioxide levels will likely be more of a threat, than a boon, to agricultural yields.

To produce the 10 megajoules of food energy (625 grams of carbohydrate) required to meet the daily energy needs in a typical adult western diet, a minimum of 600-1,200 litres of water was consumed somewhere on a farm. Bear in mind that some of the carbohydrate produced by plants is used to form the stems, leaves and roots of the plants – which are often inedible for us – as well as the edible parts of the food crop (the fruits, tubers, seeds, etc). Furthermore, a large amount of the grain produced is fed to livestock to be converted into meat (which contains only 10-50% of the energy of the grain feed).

Overall, production of the food that we eat each day typically requires 2,000-5,000 litres of water (depending upon the particular crop, climate, agricultural methods, and meat content in the diet). This is about one-thousand times greater than the amount of water we drink, and at least ten times greater than the total water used by an average person for washing, cooking, toilet flushing and other purposes.

Growing food and other crops accounts for most of the fresh water used by humanity. This water is provided either by rainfall falling directly on crops, or rainfall run-off that is collected in dams, lakes and rivers, or rainfall that seeps into the soil and collects within porous rocks below the surface. Some of these groundwater deposits have accumulated very slowly over long periods of time. "Fossil ground water" may be hundreds or thousands of years old.

Providing the huge amounts of water required to grow crops poses a major challenge (perhaps *the* major challenge) for expanding food production to meet the needs of the world's growing population. Many areas of the world are already experiencing "water stress", and this will likely be exacerbated by shifting weather patterns due to climate change.

In many areas of the world, agricultural yields are enhanced by irrigating crops with ground water. But many sources of ground water are being depleted as water is withdrawn at a far greater rate than it is replenished by rain. Reliance on non-renewable groundwater is not a long-term sustainable strategy, and eventually agricultural production will be limited by the amount of rain that falls throughout the year. Annual rainfall levels vary widely across the Earth, from about 0.2 metres for arid grasslands, to 1.0 metre for typical cropland, to several metres for monsoonal tropical areas.

There is considerable scope to improve the water efficiency of growing crops through improved agricultural practice, but a minimum water consumption arises from the interchange of carbon dioxide gas and water vapour across the membranes within plant leaves. I have often wondered why evolution and natural selection has not "solved this problem" by developing cell membranes which can selectively block the loss of water vapour while allowing carbon dioxide molecules to enter. I wonder if it might be possible for scientists to develop plants with this ability by applying genetic engineering or selective breeding techniques.

These are critical issues. They impact on the prospect for agricultural scientists to circumvent the nexus between plant growth and water consumption.

There is no fundamental principle that would prevent cell membranes from selectively controlling the passage of carbon dioxide and water vapour.

It could be argued that the development of such an ability through evolution would have conferred a huge survival advantage in arid landscapes. If a genetic mutation enabled plant membranes to even partially block the loss of water molecules, this trait should have quickly proliferated among the following generations. The fact that this trait did not become established during the 570 million years of evolution of land-based plants suggests that it might simply not be possible. Perhaps water and carbon dioxide molecules are so similar in size and shape that cell membranes cannot distinguish between them. Or perhaps, water molecules comprise such an integral part of the molecular-scale structure of cell membranes. Or maybe, chemical processes that could actively block the passage of water molecules, while freely admitting carbon dioxide molecules, are too complex and require too much energy to be expended by plant cells. Then, perhaps the reason why a half billion years of evolution failed to "solve the problem" of selectively blocking the passage of water molecules through their membranes is because no satisfactory solution exists. If this is the case, then attempts by scientists to breed or engineer plants with this trait seem doomed to failure.

On the other hand, it is possible that plants have not developed the ability to selectively control the passage of water and carbon dioxide because they didn't need to. Maybe the reason that plants haven't "solved this problem" is because it is **our problem**, not theirs. We humans would like crop plants to produce the maximum edible carbohydrate with the minimum amount of water, but maybe this ability is not important for plants to survive and reproduce in the natural world.

Most plants that grow in deserts or arid regions are opportunistic. During long periods without rainfall, these plants close their stomata and stop producing carbohydrate. They simply survive, biding their time in a dormant state, waiting patiently for the next rainfall. When rain does come, these plants take full advantage of the available water to grow and reproduce. For these plants, being able to grow with little water may confer little survival advantage if no water is available for months or years.

Evaporation in the leaves causes water to be transported from the soil, through the plant roots and up through the stem of the plant. This water carries dissolved nutrients (like nitrate, phosphate and potassium ions) from the soil to the leaves, and these nutrients are essential to make protein and DNA. If plant growth is limited by low nutrient levels, evaporation of large quantities of water from the leaves would be beneficial in transporting nutrients that the plant needs.

Furthermore, evaporation of water cools plant leaves and would help prevent overheating (since, after all, leaves are solar collectors), and thus, would assist plants to maintain optimal temperature conditions for growth.

Thus, perhaps the ability of plants to block the loss of water vapour did not arise simply because it was not necessary or advantageous for the plant's survival. Perhaps this ability could be developed by human intervention to assist *our* long-term survival.

Notes

1. Availability of water is not always the limiting factor for plant growth. In most rainforest environments, ample water is often available but soils are nutrient-deficient. Here, plant growth is generally limited by the availability of essential nutrients like magnesium (needed to produce chlorophyll), nitrogen (needed to make protein) or phosphorus (needed to make DNA).

2. The mass of a molecule can be determined by adding the atomic masses of all the atoms it contains. Water has a molecular mass of 18; carbon dioxide has a molecular mass of 44.

    Molecules in a gas have the same average kinetic energy, determined only by the temperature. The average kinetic energy of a gas molecule is given by its molecular mass and the square of its average molecular velocity. The molecular mass of carbon dioxide is 2.4 times (44/18 = 2.4) that of water, so the ratio of their molecular velocities is given by the square root of 2.4. Consequently, the average velocity of water vapour molecules is greater by a factor of 1.56.

About the author

Martin Gellender has always been fascinated by science. In his native New York, he graduated with a bachelor's degree in Chemistry, worked for a major pharmaceutical company, and did a PhD in Physical Chemistry (at the City University of New York). He moved to Canada, where he worked as a science writer for a chemical engineering magazine (and married an Aussie), and then moved to England to start a chemistry magazine.

In 1982, Marty relocated to Brisbane and spent most of his career in the Queensland Government. He played a key role in setting up an Energy Information Centre, and in negotiating an agreement between the Queensland Government and CSIRO to establish the Queensland Centre for Advanced Technology (QCAT). Most recently, he managed a grants program that funded companies developing energy-efficient and water-saving technology.

Marty is still trying to understand how the world works (attending U3A and on-line courses, and various lectures). He takes great pleasure sharing what he has learned with his four grand-kids, students in his course "Science and the Big Issues of Our Time" at the University of the Third Age (U3A), guiding visitors around Brisbane as a Brisbane Greeter, and in writing this book.

www.ingramcontent.com/pod-product-compliance
Ingram Content Group UK Ltd.
Pitfield, Milton Keynes, MK11 3LW, UK
UKHW061212180426
11947UKWH00028B/2007